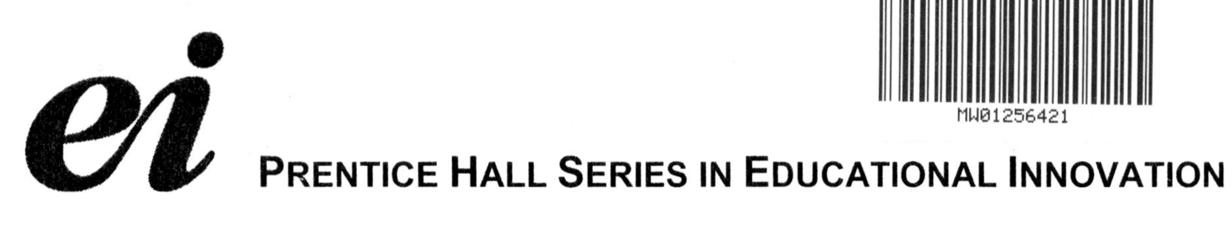

E & M TIPERs

Electricity and Magnetism Tasks
(Inspired by Physics Education Research)

Curtis J. Hieggelke
Joliet Junior College

David P. Maloney
Indiana University Purdue University Fort Wayne

Stephen E. Kanim
New Mexico State University

Thomas L. O'Kuma
Lee College

Upper Saddle River, NJ 07458

Senior Editor: Erik Fahlgren
Associate Editor: Christian Botting
Executive Managing Editor: Kathleen Schiaparelli
Production Editor: Shari Toron
Manufacturing Manager: Alexis Heydt-Long
Manufacturing Buyer: Alan Fischer
Senior Marketing Manager: Shari Meffert
Editorial Assistant: Jessica Berta
Copy Editor: Karen Bosch
Art Director: Jayne Conte
Cover Designer: Bruce Kenselaar

Pearson Prentice Hall
Pearson Education, Inc.
Upper Saddle River, NJ 0745

Printed in the United States of America

10 9 8 7 6 5 4 3

ISBN 0-13-185499-2

Pearson Education Ltd., *London*
Pearson Education Australia Pty., Limited, *Sydney*
Pearson Education Singapore, Pte. Ltd.
Pearson Education North Asia Ltd., *Hong Kong*
Pearson Education Canada, Ltd., *Toronto*
Pearson Educación de Mexico, S.A. de C.V.
Pearson Education—Japan, *Tokyo*
Pearson Education Malaysia, Pte. Ltd.

Prentice Hall Series in Educational Innovation

Lecture Tutorials for Introductory Astronomy
Jeffrey P. Adams
Edward E. Prather
Timothy F. Slater
Jack Dostal
The Conceptual Astronomy and Physics Education Research (CAPER) *Team*

Physlet® Quantum Physics: An Interactive Introduction
Mario Belloni
Wolfgang Christian
Anne J. Cox

Physlet® Physics: Interactive Illustrations, Explorations and Problems for Introductory Physics
Wolfgang Christian
Mario Belloni

Physlets®: Teaching Physics with Interactive Curricular Material
Wolfgang Christian
Mario Belloni

Peer Instruction for Astronomy
Paul J. Green

Peer Instruction: A User's Manual
Eric Mazur

Tutorials in Introductory Physics
Lillian C. McDermott
Peter S. Shaffer
The Physics Education Group, University of Washington

Just-In-Time Teaching: Blending Active Learning with Web Technology
Gregor M. Novak
Evelyn T. Patterson
Andrew D. Gavrin
Wolfgang Christian

Ranking Task Exercises in Physics: Student Edition
Thomas L. O'Kuma
David P. Maloney
Curtis J. Hieggelke

Learner-Centered Astronomy Teaching: Strategies for ASTRO 101
Timothy F. Slater
Jeffrey P. Adams

ABOUT THE AUTHORS

Since 1991, Curtis Hieggelke, David Maloney, Stephen Kanim, and Thomas O'Kuma have led over 30 workshops in which educators learned how to use and develop TIPERs. Many of these workshops were part of the Two Year College PhysicsWorkshop Project (supported by seven grants from the National Science Foundation and directed by Curtis Hieggelke and Thomas O'Kuma), which has offered a series of more than 60 professional development workshops for over 1200 Two-Year College and High School physics teacher participants. Working with Alan Van Heuvelen, they also developed the Conceptual Survey of Electricity and Magnetism, which has become a standard instrument for measuring electricity and magnetism conceptual gains in introductory physics courses.

Curtis J. Hieggelke received a B.A. in physics and mathematics from Concordia College and a Ph.D in theoretical particle physics from the University of Nebraska. He spent his professional career as a physics teacher at Joliet Junior College until he retired in 2003. Hieggelke has served as President and Section Representative of the Illinois Section of the AAPT, and received the Distinguished Service Citation in 1993. Joliet Junior College received the Illinois Community College Board Excellence in Teaching/Learning Award for his work in 1992. He was awarded the Distinguished Service Citation by AAPT in 1994 and he was elected to the Executive Board of AAPT for three years as the Two-Year College Representative. He has received numerous NSF grants for workshops and curriculum materials development based on PER.

David P. Maloney received a B.S. in physics from the University of Louisville, and a M.S. in physics and Ph.D. in physics, geology and education from Ohio University. A member of the Indiana University-Purdue University-Fort Wayne faculty since 1987, he has also taught at Wesleyan College and Creighton University. He was awarded the Distinguished Service Citation by AAPT in 2005. His main research interests concern the study of students' common sense ideas about physics, how those ideas interact with physics instruction, and the study of problem solving in physics. Maloney has authored or co-authored two-dozen articles and has been the Principal Investigator or Co-Principal Investigator for eight National Science Foundation (NSF) grant projects.

Stephen E. Kanim received a B.S. in electrical engineering from UCLA (University of California, Los Angeles) and a Ph.D. in physics from the University of Washington. He is a member of the physics faculty at New Mexico State University, where he conducts research into student's conceptual development in physics. He previously taught high school physics in Las Cruces, New Mexico and in Palo Alto, California, and worked as an electrical engineer in Santa Clara, California. He is currently developing laboratory materials for introductory mechanics with support from NSF.

Thomas L. O'Kuma received his B.S. and M.S. in physics and mathematics from Louisiana Tech University. Since 1976 he has been a full-time, two-year college physics instructor, and has been at Lee College since 1989. He wrote the *Study Guide to Beiser's Physics* and has written various lab manuals for the introductory physics course. He received the Distinguished Service Citation in 1994 and the Award for Excellence in Introductory College Physics Teaching in 2002 from the American Association of Physics Teachers (AAPT), and the Robert N. Little Award in 1994 for Outstanding Contributions to Physics in Higher Education in Texas. O'Kuma's research interests are in how students learn physics and in developing tools to help students learn physics.

Contents

Electrostatics

Magnetism

PREFACE

This workbook is intended to improve understanding of some of the ideas underlying electricity and magnetism. As the subtitle of the workbook *Tasks Inspired by Physics Education Research* (TIPERs) suggests, the design of the individual tasks within the workbook is based on research that has been conducted into how students learn physics. This research has focused on students who are facing the challenges of an introductory physics course. Through interviews with these students and through analyses of their responses to examination and homework questions, physics education researchers are developing a better understanding of common difficulties with physics, and of the types of exercises that foster improved understanding. In some cases, the tasks in this workbook are directly based on questions that researchers have used to probe student understanding. In other cases, we have written exercises that are not directly taken from research, but that focus on concepts that research suggests are challenging to students.

One general finding of physics education research is that many students in introductory physics courses who attend lectures, read the textbooks, and solve the homework problems still struggle with important physics concepts, principles, and relationships. This difficulty is sometimes a result of conflicts between common sense ideas about how nature behaves (ideas based on everyday experiences) and the physics rules being learned (ideas based on rigorous investigation and controlled experimentation). Some of the tasks we have included in this workbook are intended to focus on these conflicts between everyday experience and how physicists think about concepts. By directly comparing these conflicting ways of thinking about specific concepts, and by discussing them with peers, we expect that students will clarify which of their initial ideas are useful and which need to be modified.

A second source of difficulty for students is that understanding some of the topics in electricity and magnetism requires application of concepts and principles that were introduced in mechanics. Often, these concepts were only partially understood when they were introduced, and then become even more difficult in a new context. We have included tasks in this workbook that are designed to reinforce mechanics concepts and relationships that physics education researchers have identified as challenging for many students. Still other tasks are based on the recognition that some concepts in physics require repeated exposure in a variety of contexts to really "sink in."

First and foremost one should understand that this is *not a book that is just read to get some needed information*. To benefit from this workbook, an individual must **actively** work with the tasks. The tasks in this book focus on conceptual understanding. Consequently, plugging numbers into a formula will seldom be the way to solve these tasks. Looking up an answer in a physics book is also *not* the way to deal with these tasks. Completing the tasks and then discussing answers with others is a useful way to benefit from this book. The more someone talks about these ideas and issues, the better the concepts will be learned and understood.

The tasks in this workbook will help deepen an individual's understanding of electricity and magnetism. Beyond this, we would like these tasks to reinforce the sense that the *ideas* of science have coherence and power that extend beyond the facts and equations.

There are eleven types of task formats in this book. These formats are likely to be new to most people so those people will require a little time to learn each format, *i.e.*, how the information is presented, and what someone needs to do to solve the task. The formats are reasonably straightforward, so getting familiar with them should not require significant time or effort. However, many of the tasks ask for explanations of the work or the reasoning associated with reaching the answer. This is one of the most important parts of these tasks. The explanation is especially important for cases where the correct answer is "none of them" or "it cannot be determined."

The different formats are designated by letters identifying the task type according to the following system: Bar Chart Tasks—BCT, Changing Representations Tasks—CRT, Comparison Tasks—CT,

Conflicting Contentions Task—CCT, Linked Multiple Choice Tasks—LMCT, Predict and Explain Tasks—PET, Qualitative Reasoning Tasks—QRT, Ranking Tasks—RT, Troubleshooting Tasks—TT, What, if anything, is Wrong Task—WWT, and Working Backwards Tasks—WBT. In addition to the format identifier, each task begins with either an eT or mT which indicates an electricity task or a magnetism task. Task titles then describe the physical situation and finally identify the target quantity being asked about. It is unlikely that an instructor will go through this book in sequence; it is more likely that there will be some jumping around as an instructor chooses different formats to use at different times.

There are some unique aspects to consider for the Ranking Task format. If there are two, three, or four of the variations that have equivalent values for the target quantity, it is necessary to explicitly show that they are tied when writing the ranking sequence. For example, if choices A and C have the same ranking which is greater than B, then the answer would be AC in the first slot, followed by a blank slot, and then B would be in the last slot (or A, C, B can be put in sequence with a circle enclosing A and C). With ranking tasks it may not be possible to figure out specific numerical values for a quantity, but it may still be possible to compare the situations to decide which is largest and so on. Consequently, it is possible to rank the situations.

Several common conventions are employed in the tasks in this book. A circle with a dot in the center is used to represent a vector pointing out of the page, and a circle with an x in the center is used to represent a vector pointing into the page. Unless clearly stated, all grids have the same spacing and related drawings have the same scale. The exception will be when the term "far apart" is used since the page size limits the drawing size. Uniform fields, electric or magnetic, will be constant both in space and in time unless otherwise specified. We ignore gravity and friction in these tasks unless explicitly identified. Electric potentials for point charges are zero far away from the charges.

ACKNOWLEDGMENTS

An endeavor like this book requires input from many people in addition to the authors. We sincerely thank the following individuals. At Joliet Junior College (Joliet, IL): President J. D. Ross, Vice President for Academic Affairs Dr. Denis Wright, Judy Bucciferro, Dr. Max Lee, Natalie Ward, Geoff White, Christi Wren, and current and former students Leeanne Daoust, Gustavo A. Gil Jr., Carl Guzman, Brian Hawkins, Blake Johnson, Brian Ruddy, and Michael Schneidewind. At Lee College (Baytown, TX): Regina Barrera, William Kominek, and Kathy O'Kuma. Also Vince Garcia at IPFW (Fort Wayne, IN) and Todd Leif of Cloud County Community College (Concordia, KS). These tasks were extensively reviewed and improved by Martha Lietz of Niles West High School (Skokie, IL), Dr. Robert Morse of St. Albans School (Washington, DC), and Dr. William P. Hogan (Joliet Junior College, IL). We also valued the feedback from the many participants who have attended our workshops.

Since this book contains Tasks Inspired by Physics Education Research, we would like to acknowledge several of our most important inspirations. First, Alan Van Heuvelen deserves mention because his ALPS manual, which used bar charts and changing representation tasks, provided an early and strong guiding idea, and he was the co-developer of the Working Backwards (Jeopardy) format. Second, the University of Washington Physics Education Group frequently used the Conflicting Contentions format and provided ideas for issues in electricity and magnetism. Third, we thank the many PER investigators whose names we unfortunately cannot mention individually. Finally, we want to recognize and thank Paul G. Hewitt for his pioneering work in promoting conceptual understanding in physics courses.

We would like to acknowledge and thank the National Science Foundation (DUE #9952735 and #0125831) and Duncan McBride in particular, for supporting the development of these materials and the Physics Workshop Projects. Finally, we thank Erik Fahlgren, Christian Botting, and the staff at Prentice-Hall for their assistance and their willingness to publish a unique type of book.

ET1-RT1: CHARGED INSULATING BLOCKS—CHARGE DENSITY

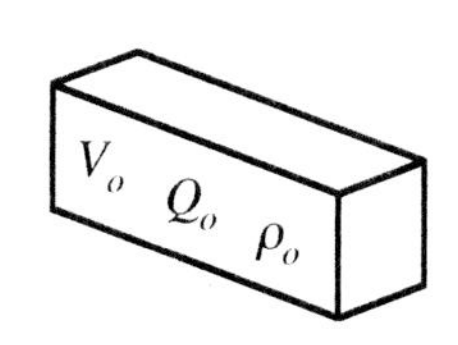

The block of insulating material shown at right has a volume V_o. An overall charge Q_o is spread evenly throughout the volume of the block so that the block has a uniform charge density ρ_o.

Six additional charged insulating blocks are shown below. For each block, the volume is given as well as *either* the charge or the charge density.

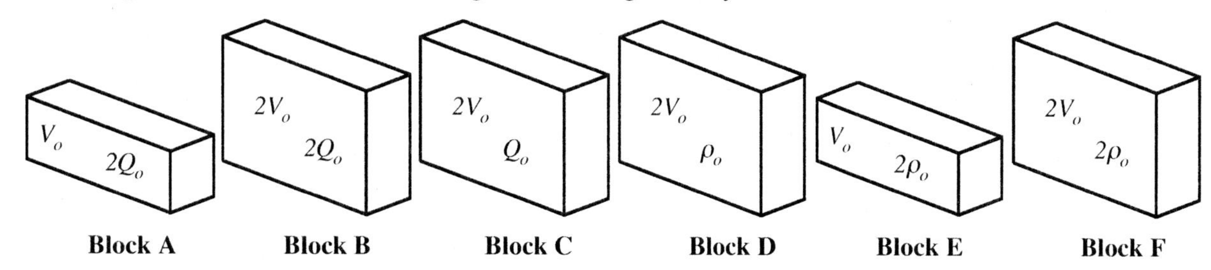

Rank the charge densities of the six blocks.

Greatest 1 ________ 2 ________ 3 ________ 4 ________ 5 ________ 6 ________ Least

OR, the charge density is the same for all six blocks. ____

OR, the ranking for the charge density cannot be determined. ____

Carefully explain your reasoning.

How sure were you of your ranking? (circle one)

Basically Guessed					Sure				Very Sure
1	2	3	4	5	6	7	8	9	10

ET1-RT2: BREAKING A CHARGED INSULATING BLOCK—CHARGE DENSITY

A block of insulating material (labeled O in the diagram) with a width w, height h, and thickness t has a positive charge $+Q_o$ distributed uniformly throughout its volume. The block is then broken into three pieces, A, B, and C, as shown.

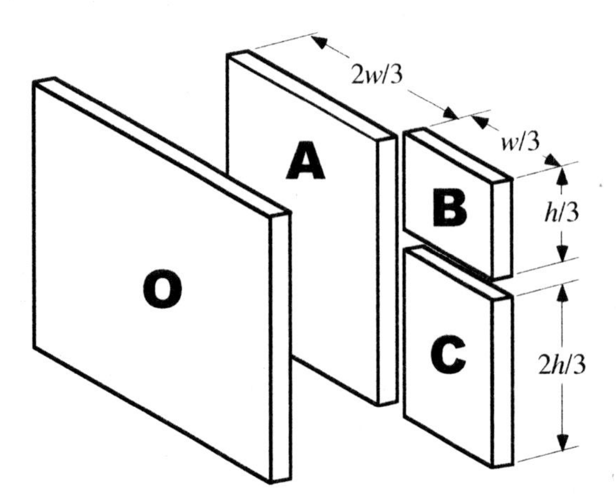

Rank the charge densities of the original block O, piece A, piece B, and piece C.

Greatest 1 _____ 2 _____ 3 _____ 4 _____ Least

OR, the charge density is the same for all four pieces. ____

OR, the ranking for the charge densities cannot be determined. _____

Carefully explain your reasoning.

How sure were you of your ranking? (circle one)

Basically Guessed					Sure				Very Sure
1	2	3	4	5	6	7	8	9	10

ET1-RT3: CHARGED INSULATING BLOCKS—CHARGE

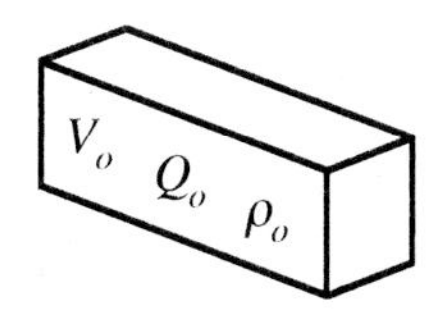

The block of insulating material shown at right has a volume V_o. An overall charge Q_o is spread uniformly throughout the volume of the block so that the block has a charge density ρ_o.

Six additional charged insulating blocks are shown below. For each block, the volume is given as well as *either* the charge or the charge density of the block.

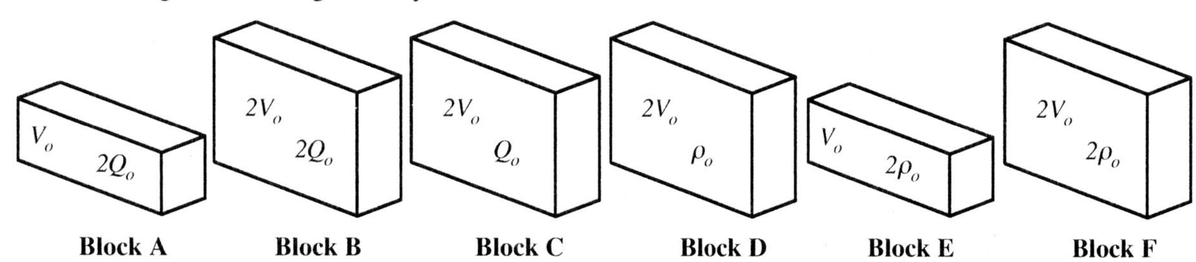

Rank the overall charge of the six blocks.

Greatest 1 ______ 2 ______ 3 ______ 4 ______ 5 ______ 6 ______ Least

OR, the charge is the same for all six blocks. ____

OR, the ranking for the charge cannot be determined. ____

Carefully explain your reasoning.

How sure were you of your ranking? (circle one)

Basically Guessed				Sure					Very Sure
1	2	3	4	5	6	7	8	9	10

ET1-RT4: PAIRS OF CONNECTED CHARGED CONDUCTORS—CHARGE

Three pairs of charged, isolated, conducting spheres are connected with wires and switches. The spheres are very far apart. The large spheres have twice the radius of the small spheres. Each sphere on the left has a charge of +20 nC and each sphere on the right has a charge of +70 nC before the switches are closed.

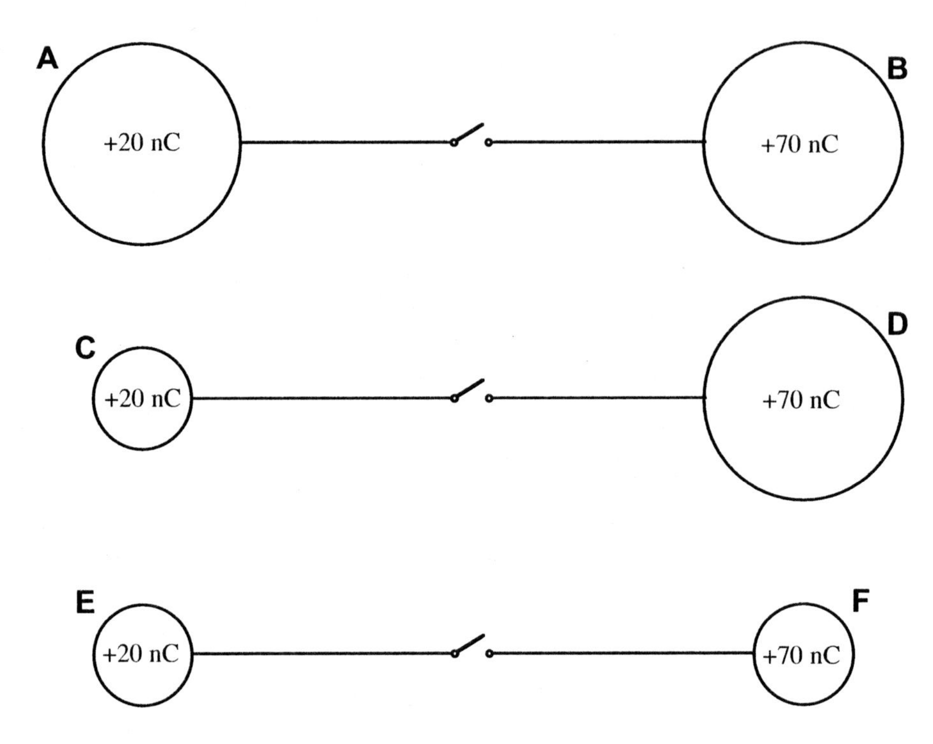

Rank the electric charge of the spheres after all of the switches are closed.

Greatest 1 _______ 2 _______ 3 _______ 4 _______ 5 _______ 6 _______ Least

OR, the electric charge is the same for all six spheres. _____

OR, the ranking of the electric charge cannot be determined. _____

Carefully explain your reasoning.

How sure were you of your ranking? (circle one)

Basically Guessed				Sure					Very Sure
1	2	3	4	5	6	7	8	9	10

ET1-RT5: COLLECTION OF SIX CHARGED CONNECTED CONDUCTORS—CHARGE

Six charged conducting spheres are connected with wires and switches. The large spheres have twice the radius of the small spheres. Each sphere on the left has a charge of +20 nC and each sphere on the right has a charge of +70 nC before the switches are closed.

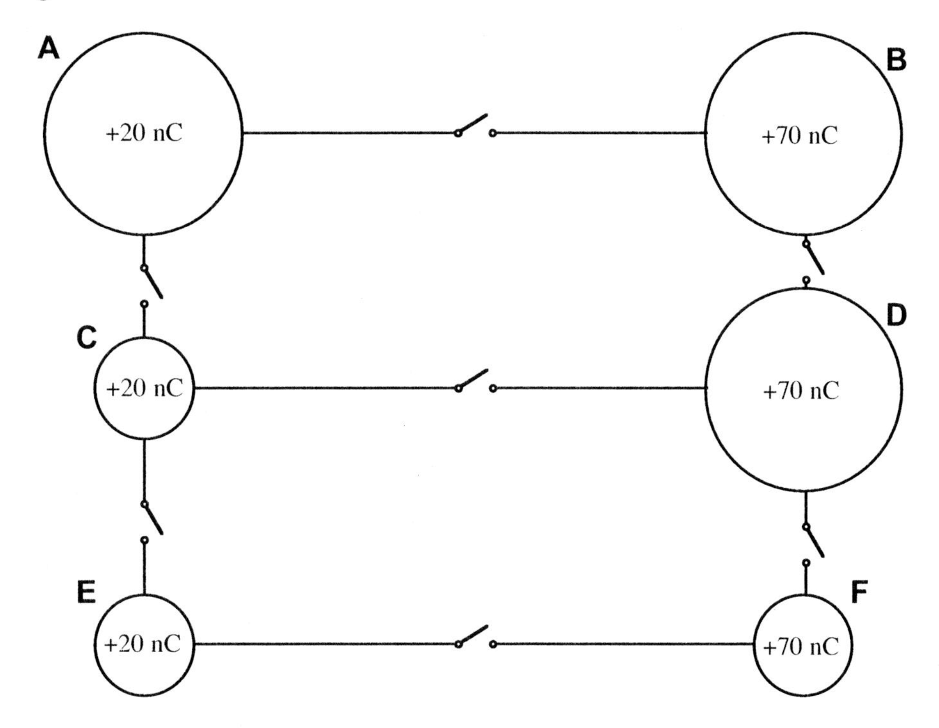

Rank the electric charge of the spheres after all of the switches are closed.

Greatest 1 _______ 2 _______ 3 _______ 4 _______ 5 _______ 6 _______ Least

OR, the electric charge is the same for all six spheres. _____

OR, the ranking of the electric charge cannot be determined. _____

Carefully explain your reasoning.

How sure were you of your ranking? (circle one)

Basically Guessed / Sure / Very Sure

1 2 3 4 5 6 7 8 9 10

ET1-RT6: PAIRS OF OUTSIDE AND INSIDE CONNECTED CHARGED CONDUCTORS—CHARGE

Two pairs of charged, hollow, spherical conducting shells are connected with wires and switches. The system AB is very far from CD. The large shells have four times the radius of the small shells. Each pair has a charge of +20 nC on the small shell and +60 nC on the large shell before the switches are closed.

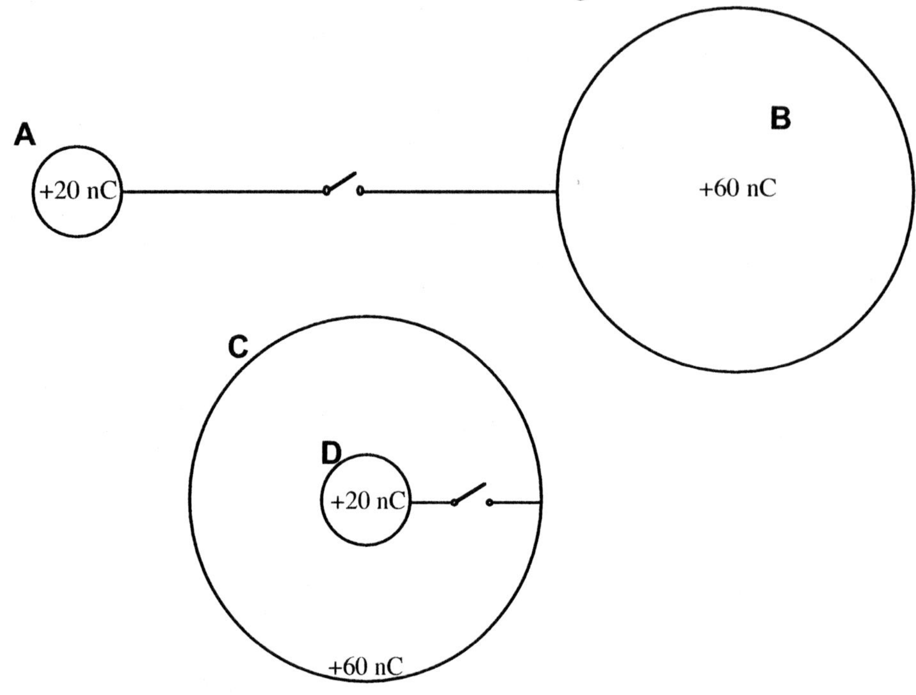

Rank the electric charge on the shells A-D after the switches are closed.

Greatest 1 _______ 2 _______ 3 _______ 4 _______ Least

OR, the electric charge is the same for all four shells. _____

OR, the ranking of the electric charge cannot be determined. _____

Carefully explain your reasoning.

How sure were you of your ranking? (circle one)

Basically Guessed				Sure					Very Sure
1	2	3	4	5	6	7	8	9	10

ET1-RT7: CHARGED ROD AND ELECTROSCOPE—EXCESS CHARGE

In each of the four cases below, a charged rod is brought close to an electroscope that is initially uncharged. In cases A and B, the rod is positively charged; in cases C and D, the rod is negatively charged. In cases A and C, the leaf of the electroscope is deflected the same amount, which is more than it is deflected in cases B and D.

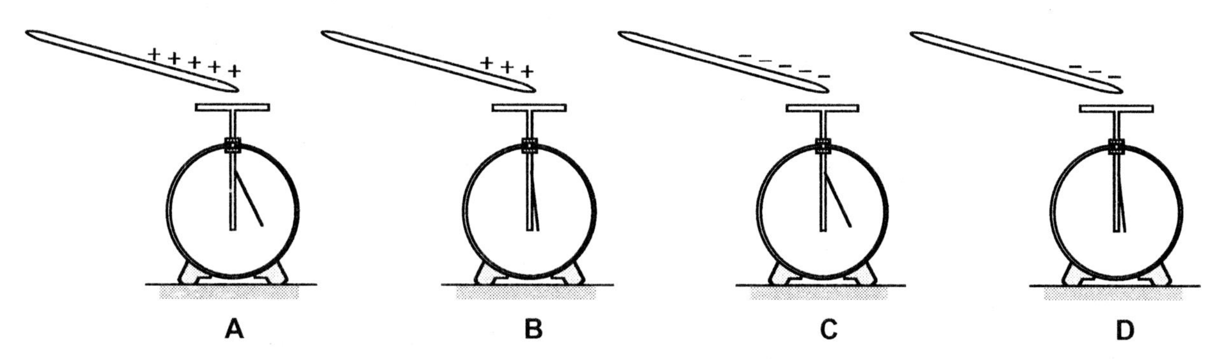

Rank the net charge on the electroscope while the charged rod is near. (This will be a negative value if there is more negative than positive charge on the electroscope.)

Greatest positive 1 _______ 2 _______ 3 _______ 4 _______ Greatest negative

OR, the net charge is the same for all four situations but it is not zero. _______

OR, the net charge is zero for all of these situations. _______

OR, the ranking for the net charge cannot be determined from the information given. _______

Carefully explain your reasoning.

How sure were you of your ranking? (circle one)

Basically Guessed					Sure				Very Sure
1	2	3	4	5	6	7	8	9	10

ET3-RT1: THREE-DIMENSIONAL LOCATIONS IN A CONSTANT ELECTRIC POTENTIAL—FORCE

The electric potential has a constant value of six volts everywhere in a three-dimensional region, part of which is shown below.

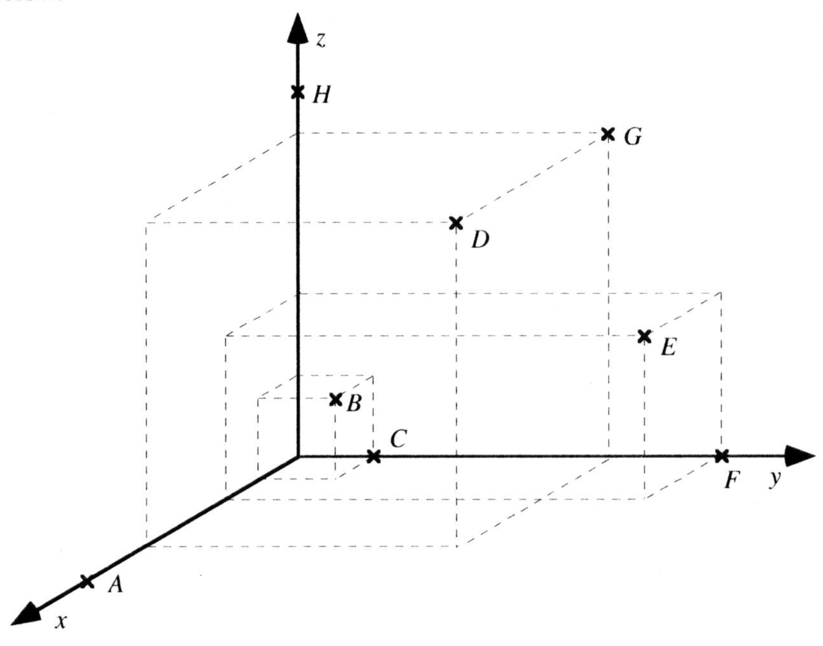

Rank the strength (magnitude) of the electric force on a charge of +2 μC if it is placed at the labeled points.

Greatest 1 ______ 2 ______ 3 ______ 4 ______ 5 ______ 6 ______ 7 ______ 8 ______ Least

OR, the electric force is the same but not zero for all of these points. ____

OR, the electric force is zero for all of these points. ____

OR, the ranking for the electric force cannot be determined for all of these points. ____

Carefully explain your reasoning.

How sure were you of your ranking? (circle one)

Basically Guessed | Sure | Very Sure

1 2 3 4 5 6 7 8 9 10

ET3-RT2: CHARGES ARRANGED IN A TRIANGLE—FORCE

In each case below, three particles are fixed in place at the vertices of an equilateral triangle. The triangles are all the same size. The particles are charged as shown. (In case C, the top particle has no charge.)

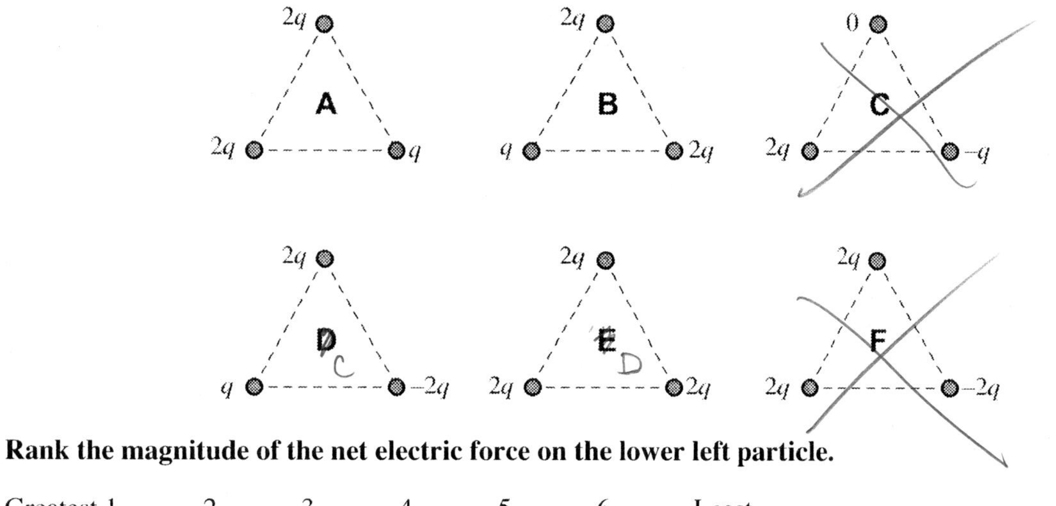

Rank the magnitude of the net electric force on the lower left particle.

Greatest 1 ______ 2 ______ 3 ______ 4 ______ 5 ______ 6 ______ Least

OR, the net electric force on the lower left particle is the same for all six cases. _____

OR, the ranking for the net electric force on the lower left particle cannot be determined. _____

Carefully explain your reasoning.

How sure were you of your ranking? (circle one)

Basically Guessed					Sure				Very Sure
1	2	3	4	5	6	7	8	9	10

ET3-RT3: CHARGES IN A PLANE—FORCE

In each case shown below, small charged particles are fixed on grids having the same spacing. Each charge q is identical, and all other charges have a magnitude that is an integer multiple of Q.

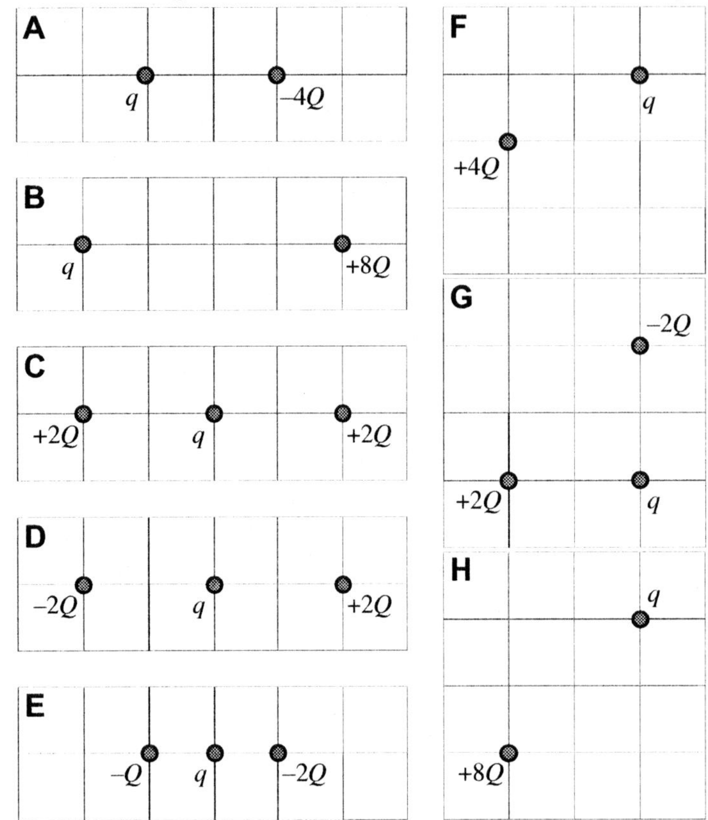

Rank the magnitude of the electric force on the charge labeled *q* due to the other charges.

Greatest 1 ______ 2 ______ 3 ______ 4 ______ 5 ______ 6 ______ 7 ______ 8 ______ Least

OR, the electric force on q is the same but not zero for all eight cases. ____

OR, the electric force on q is zero for all eight cases. ____

OR, the ranking for the electric force on q cannot be determined. ____

Carefully explain your reasoning.

How sure were you of your ranking? (circle one)

Basically Guessed				Sure					Very Sure
1	2	3	4	5	6	7	8	9	10

ET3-RT4: Two Charges—Force

In each case shown below, small charged particles are fixed on grids having the same spacing. Each charge q is identical, and all the other charges have a magnitude that is an integer multiple of q.

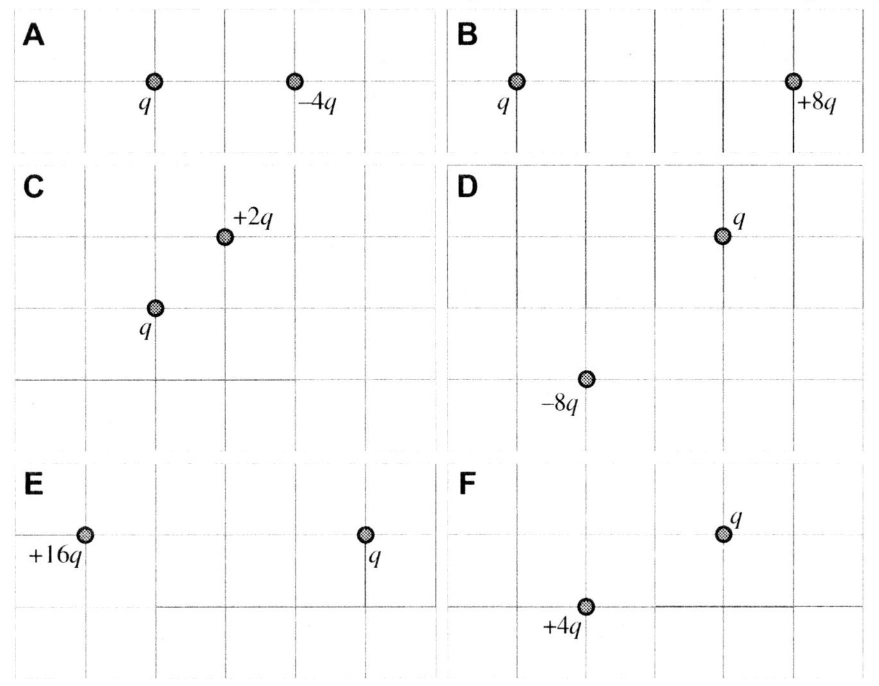

Rank the magnitude of the electric force on the charge labeled *q* due to the other charge.

Greatest 1 ______ 2 ______ 3 ______ 4 ______ 5 ______ 6 ______ Least

OR, the electric force on q is the same but not zero for all six cases. _____

OR, the electric force on q is zero for all six cases. _____

OR, the ranking for the electric force on q cannot be determined. _____

Carefully explain your reasoning.

How sure were you of your ranking? (circle one)

Basically Guessed					Sure				Very Sure
1	2	3	4	5	6	7	8	9	10

ET3-RT5: TWO AND THREE CHARGES IN A LINE—FORCE

In each case shown below, small charged particles are fixed on grids having the same spacing. Each charge q is identical, and all the other charges have a magnitude that is an integer multiple of q.

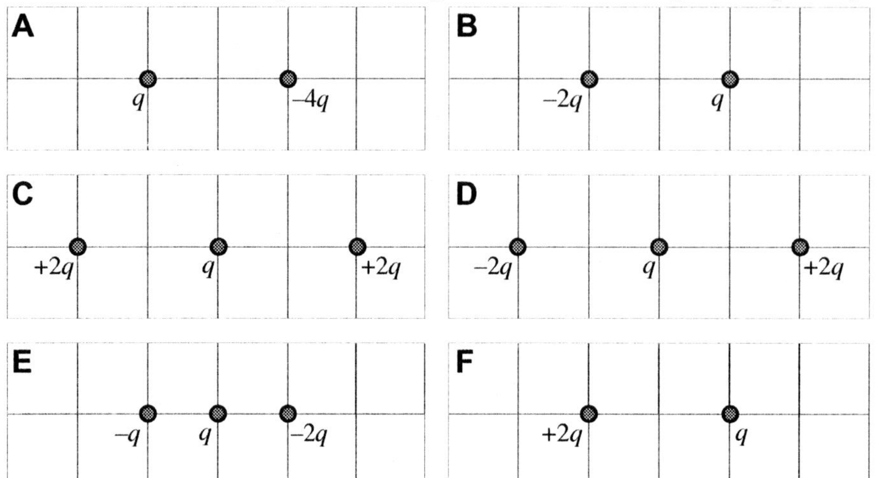

Rank the magnitude of the electric force on the charge labeled q due to the other charges.

Greatest 1 _____ 2 _____ 3 _____ 4 _____ 5 _____ 6 _____ Least

OR, the electric force on q is the same but not zero for all six cases. ____

OR, the electric force on q is zero for all six cases. ____

OR, the ranking for the electric force on q cannot be determined. ____

Carefully explain your reasoning.

How sure were you of your ranking? (circle one)

Basically Guessed					Sure				Very Sure
1	2	3	4	5	6	7	8	9	10

ET3-RT6: CHARGED RODS AND POINT CHARGES—FORCE

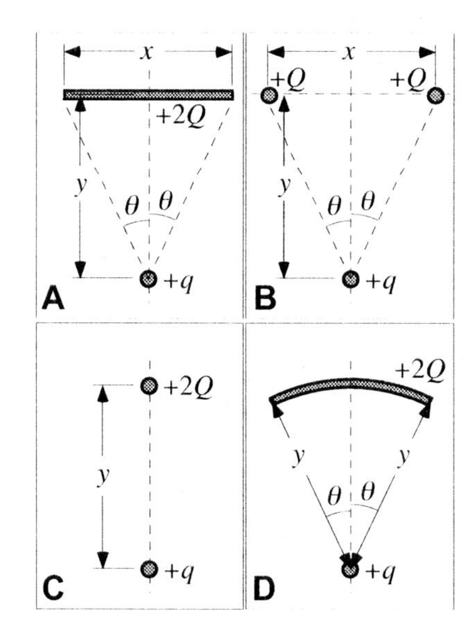

In each case A-D, a point charge $+q$ is fixed in place as well as some other point charges or charged rods.

The charged insulating rod in case A has a length x and a charge $+2Q$ distributed uniformly along it. The charged insulating rod in case D is an arc of radius y, and has a charge $+2Q$ distributed uniformly along it.

Rank the magnitude of the electric force on $+q$ due to the other charges in each case.

Greatest 1 _____ 2 _____ 3 _____ 4 _____ Least

OR, the electric force on +q is the same for all four cases. ____

OR, the ranking for the electric force on +q cannot be determined.____

Carefully explain your reasoning.

How sure were you of your ranking? (circle one)

Basically Guessed				Sure					Very Sure
1	2	3	4	5	6	7	8	9	10

ET3-RT7: CHARGED CURVED ROD—FORCE

A point charge $+q$ is placed near a curved, charged, insulating rod as shown at left below. The charge is placed at the center of curvature of the curved rod. For each of the five cases A-E, the charge density on the rod varies according to the graphs, but the total charge is the same.

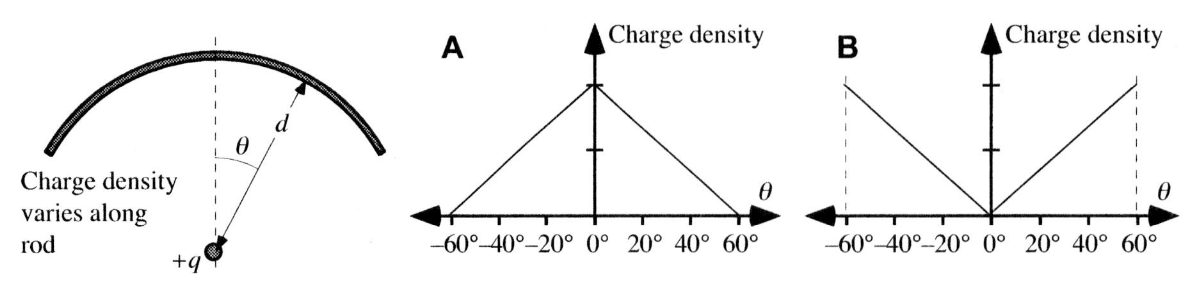

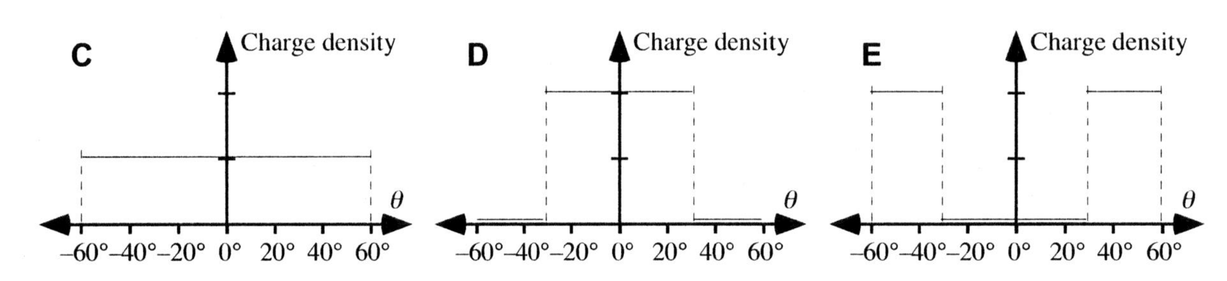

Rank the magnitude of the electric force on $+q$ due to the charge in the curved rod in each case.

Greatest 1 _______ 2 _______ 3 _______ 4 _______ 5 _______ Least

OR, the electric force on +q is the same (but not zero) for all five cases. ____

OR, the electric force on +q is zero for all five cases. ____

OR, the ranking for the electric force on +q cannot be determined. ____

Carefully explain your reasoning.

How sure were you of your ranking? (circle one)

Basically Guessed					Sure				Very Sure
1	2	3	4	5	6	7	8	9	10

ET3-RT8: THREE-DIMENSIONAL LOCATIONS NEAR A POINT CHARGE—ELECTRIC FORCE

There is a positive point charge $+q$ located at (0, 3, 0) as shown in the three-dimensional region below. Within that region are points located on the corners of two cubes as shown below. The small cube has edges of 1 centimeter length and the larger cube has edges of 3 centimeter length.

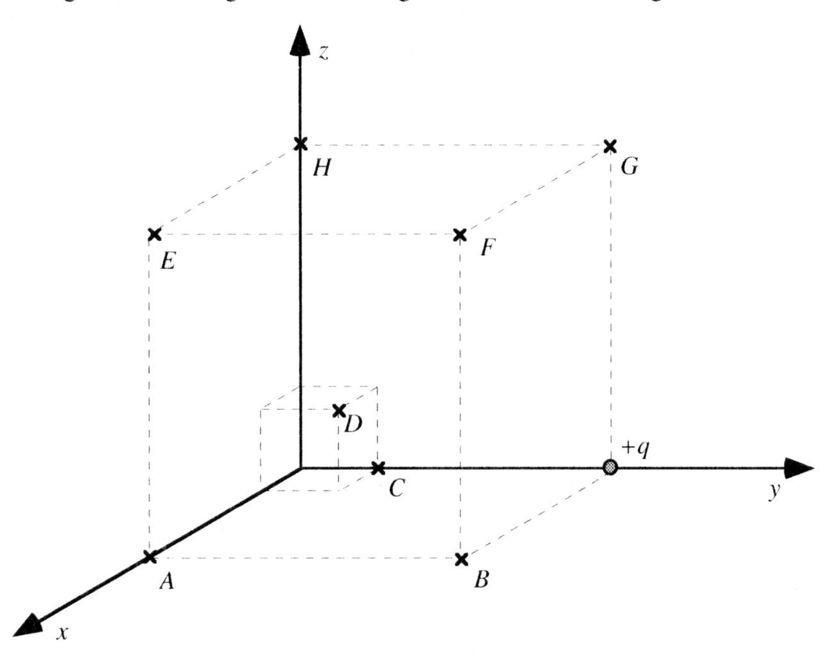

Rank the strength (magnitude) of the electric force on a $+3q$ point charge if it is placed at the labeled points.

Greatest 1 ______ 2 ______ 3 ______ 4 ______ 5 ______ 6 ______ 7 ______ 8 ______ Least

OR, the electric force is the same but not zero for all these points. _____

OR, the electric force is zero for all these points. _____

OR, the ranking for the electric force cannot be determined for all these points. _____

Carefully explain your reasoning.

How sure were you of your ranking? (circle one)

Basically Guessed — Sure — Very Sure

1 2 3 4 5 6 7 8 9 10

ET3-RT9: Sphere and a Point Charge—Force

A point charge is placed a distance d away from a neutral metal sphere. The diameters of the spheres in A and D are the same and smaller than the equal diameters in B and C. The point charge is positive for cases A and B, and negative for C and D.

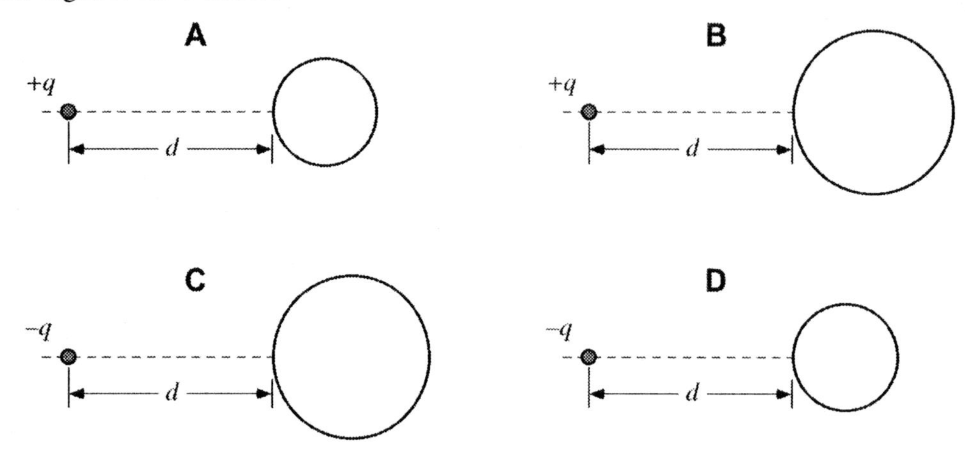

Rank the force exerted on the point charge by the sphere (let a force to the right be a positive force and a force to the left be a negative force).

Greatest positive 1 ______ 2 ______ 3 ______ 4 ______ Greatest negative

OR, the force is the same but not zero for all four situations. ____

OR, the force is zero for all these situations. ____

OR, the ranking for the forces cannot be determined. ____

Carefully explain your reasoning.

How sure were you of your ranking? (circle one)

Basically Guessed					Sure				Very Sure
1	2	3	4	5	6	7	8	9	10

ET3-RT10: Three-Dimensional Locations in a Uniform Electric Field—Electric Force

All the labeled points are within a region of space with a uniform electric field. The electric field points toward the top of the page (that is, in the positive z-direction).

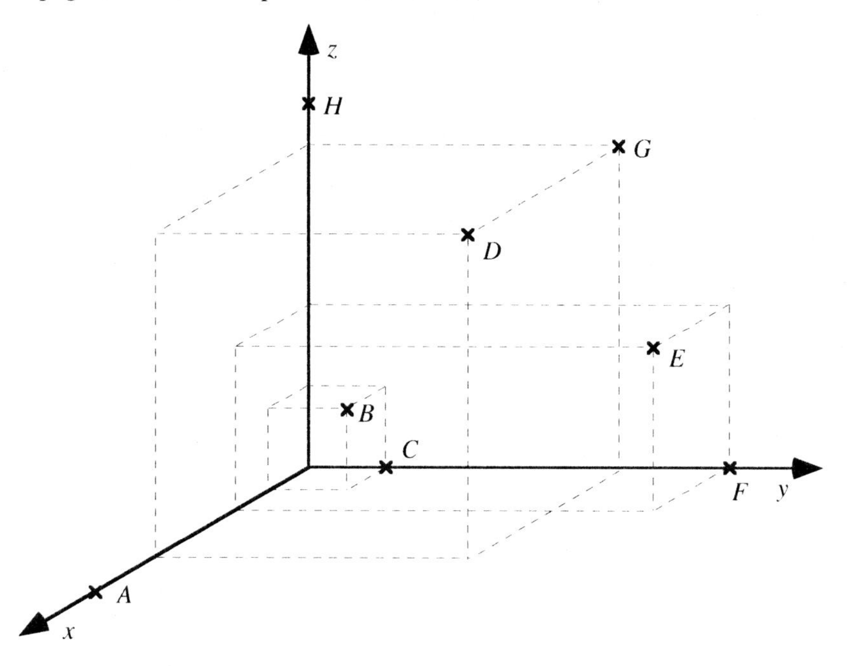

Rank the magnitude of the electric force on a charge of +2 μC at the labeled points.

Greatest 1 ______ 2 ______ 3 ______ 4 ______ 5 ______ 6 ______ 7 ______ 8 ______ Least

OR, the electric force is the same but not zero for all of these points. ____

OR, the electric force is zero for all of these points. ____

OR, the ranking for the electric force cannot be determined for all of these points. ____

Carefully explain your reasoning.

How sure were you of your ranking? (circle one)

Basically Guessed				Sure					Very Sure
1	2	3	4	5	6	7	8	9	10

ET4-RT1: TWO CHARGED OBJECTS—ACCELERATION

In each case shown below, a particle of charge $+q$ is placed a distance d from a particle of charge $+4q$. The particles are then released simultaneously. The masses of the particles are indicated in the diagram.

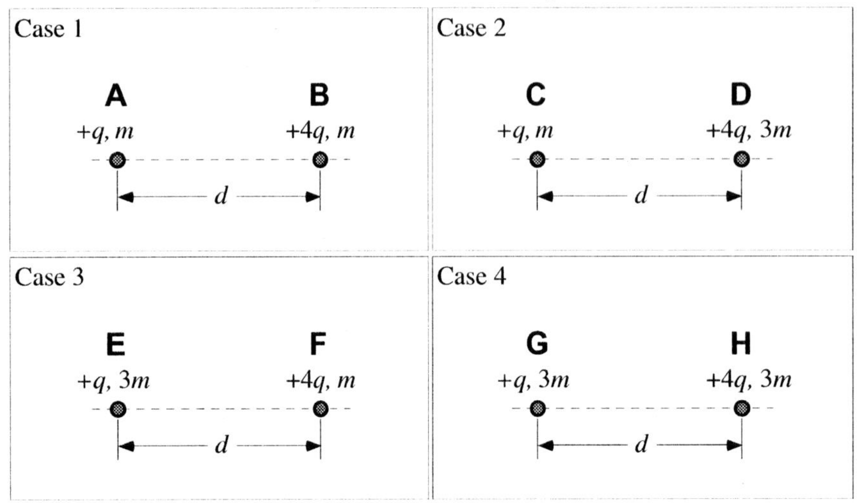

Rank the magnitude of the acceleration of each particle just after it is released.

Greatest 1 ______ 2 ______ 3 ______ 4 ______ 5 ______ 6 ______ 7 ______ 8 ______ Least

OR, the magnitude of the initial acceleration is the same but not zero for all these particles. ____

OR, the magnitude of the initial acceleration is zero for all these particles. ____

OR, the ranking for the magnitude of the initial acceleration cannot be determined. ____

Carefully explain your reasoning.

How sure were you of your ranking? (circle one)

Basically Guessed				Sure					Very Sure
1	2	3	4	5	6	7	8	9	10

ET4-RT2: CHARGES BETWEEN CHARGED PARALLEL PLATES—SPEED

Four identical positive charges are each launched with a speed v_o from a point halfway between the plates of a parallel-plate capacitor. (The plates are very large, and only a small portion near the center is shown at right.)

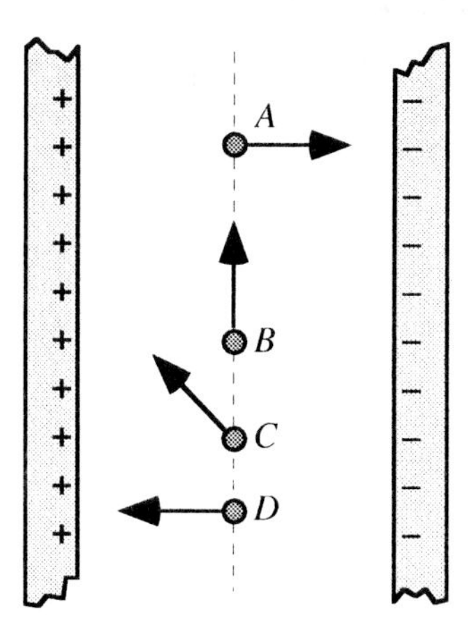

Charge *A* is launched toward the negative plate, charge *B* is launched parallel to the plates, charge *C* is launched at a 45° angle toward the positive plate, and charge *D* is launched directly toward the positive plate. Neither charge *C* nor *D* touches the positive plate.

Rank the speeds of the charges *A* – *D* just before they hit the negative plate.

Greatest 1 _______ 2 _______ 3 _______ 4 _______ Least

OR, the speed is the same for all four charges. _____

OR, the ranking for the speeds cannot be determined. _____

Carefully explain your reasoning.

How sure were you of your ranking? (circle one)

Basically Guessed					Sure				Very Sure
1	2	3	4	5	6	7	8	9	10

ET5-RT1: CHARGED INSULATING SHEETS—ELECTRIC FIELD

A distance $4d$ separates two very large, parallel insulating sheets. (Only a small portion near the center of the sheets is shown; the distance between the sheets is very small compared to the dimensions of the sheets.) The sheet on the left has a charge density +9 μC/m², and the sheet on the right has a charge density –6 μC/m².

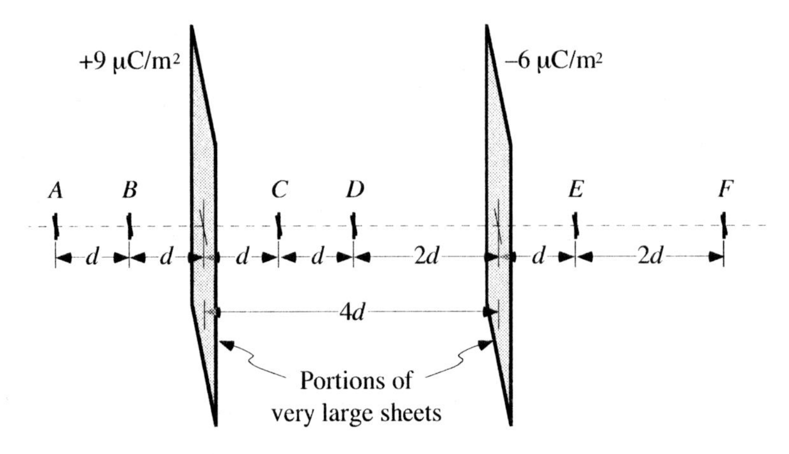

Rank the magnitude of the electric field at the points labeled *A-F*.

Greatest 1 _____ 2 _____ 3 _____ 4 _____ 5 _____ 6 _____ Least

OR, the electric field is the same but not zero for all six points. _____

OR, the electric field is zero for all six points. _____

OR, the ranking for the electric field cannot be determined. _____

Carefully explain your reasoning.

How sure were you of your ranking? (circle one)

Basically Guessed					Sure				Very Sure
1	2	3	4	5	6	7	8	9	10

ET5-RT2: CHANGING ELECTRIC FORCE ON AN ELECTRON—ELECTRIC FIELD

The graph shows the electric force acting on an electron as a function of time.

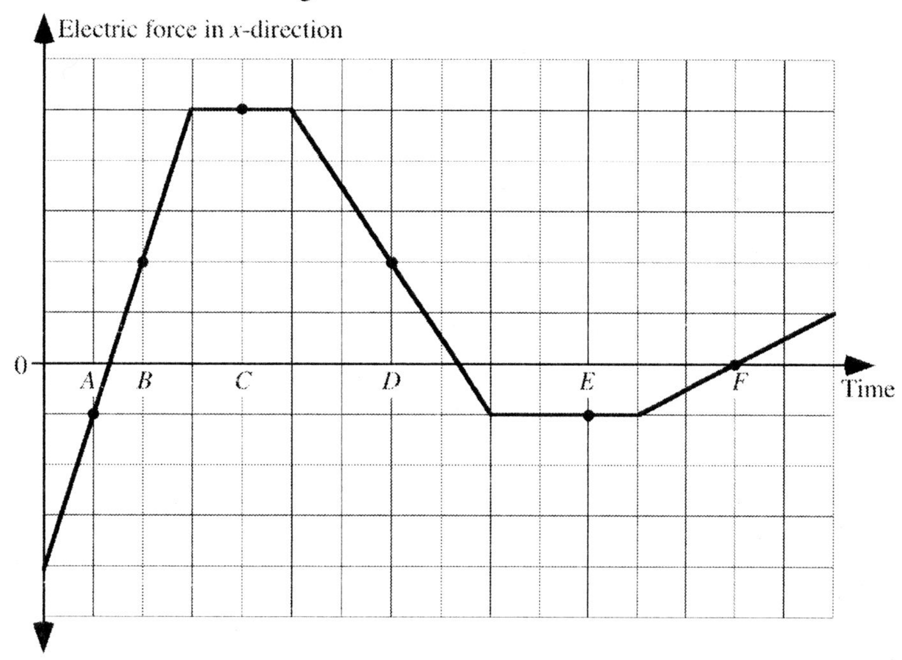

Rank the electric field at the labeled times. An electric field in the positive x-direction will be considered greater than an electric field in the negative x-direction.

Greatest Positive 1 ______ 2 ______ 3 _____ 4 _____ 5 ______ 6 ______ Greatest Negative

OR, the electric field is the same (but not zero) for all six points. _____

OR, the electric field is zero for all six of these points. _____

OR, the ranking of the electric field cannot be determined. _____

Carefully explain your reasoning.

How sure were you of your ranking? (circle one)

Basically Guessed				Sure					Very Sure
1	2	3	4	5	6	7	8	9	10

ET5-RT3: CHARGED SOLID CONDUCTING SPHERE—ELECTRIC FIELD

A solid conducting sphere with a radius of 200 cm has a charge of +300 μC.

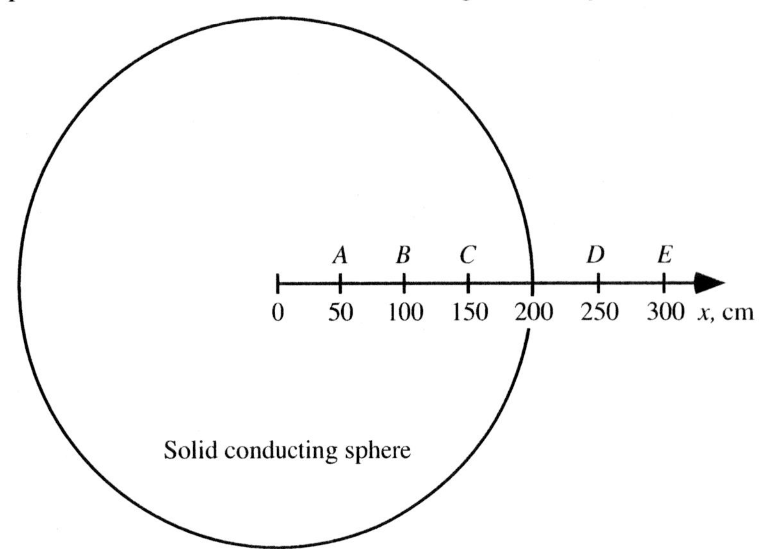

Rank the magnitude of the electric field at the labeled points.

Greatest 1________ 2________ 3________ 4________ 5________ Least

OR, the electric field is the same (but not zero) at all of the labeled points. ____

OR, the electric field is zero at all of the labeled points. ____

OR, the ranking for the electric field cannot be determined. ____

Carefully explain your reasoning.

How sure were you of your ranking? (circle one)

Basically Guessed					Sure				Very Sure
1	2	3	4	5	6	7	8	9	10

ET5-RT4: THREE-DIMENSIONAL LOCATIONS IN A CONSTANT ELECTRIC POTENTIAL—FIELD

The electric potential is a constant 8 volts everywhere in a three-dimensional region. Within that region are points located at the corners of two cubes as shown below. The small cube has edges of 1 centimeter length, and the larger cube has edges of 3 centimeters length.

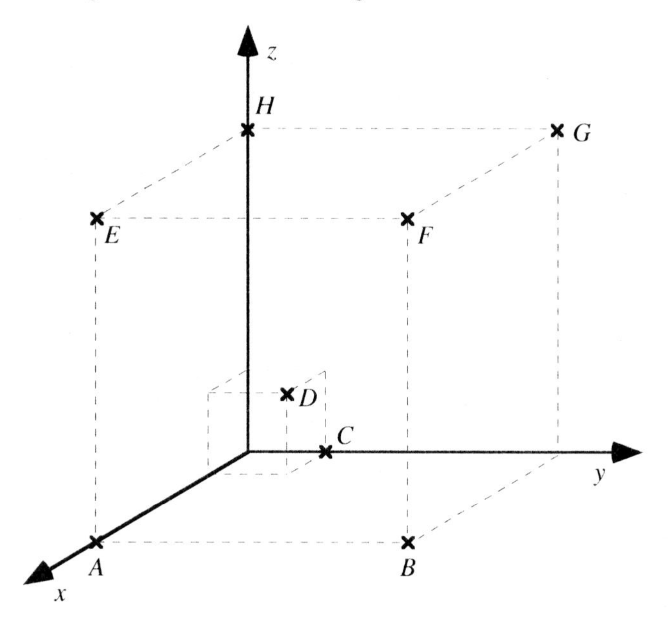

Rank the strength (magnitude) of the electric field at the labeled points.

Greatest 1 _______ 2 _______ 3 _______ 4 _______ 5 _______ 6 _______ 7 _______ 8 _______ Least

OR, the electric field is the same (but not zero) for all of these points. _____

OR, the electric field is zero for all of these points. _____

OR, the ranking for the electric field cannot be determined for these points. _____

Carefully explain your reasoning.

How sure were you of your ranking? (circle one)

Basically Guessed				Sure					Very Sure
1	2	3	4	5	6	7	8	9	10

ET5-RT5: Spherical Conducting Shell—Electric Field

A spherical conducting shell with a radius of 200 cm has a charge of +300 μC.

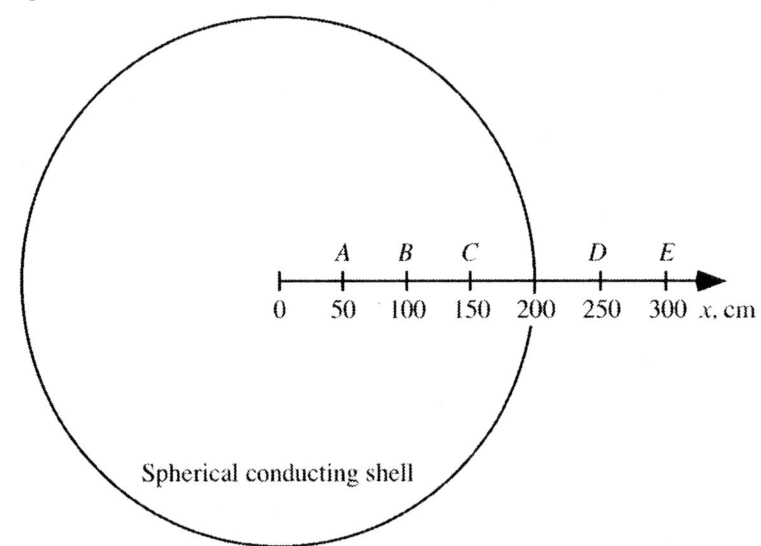

Rank the magnitude of the electric field at the labeled points.

Greatest 1_________ 2 _________ 3_________ 4_________ 5 _________ Least

OR, the electric field is the same (but not zero) at all of the labeled points. ____

OR, the electric field is zero at all of the labeled points. ____

OR, the ranking for the electric field cannot be determined. ____

Carefully explain your reasoning.

How sure were you of your ranking? (circle one)

Basically Guessed				Sure					Very Sure
1	2	3	4	5	6	7	8	9	10

ET5-RT6: Six Charges in Three Dimensions—Electric Field

In each case shown, six point charges are all the same distance away from the origin. All charges are either $+Q$ or $-Q$.

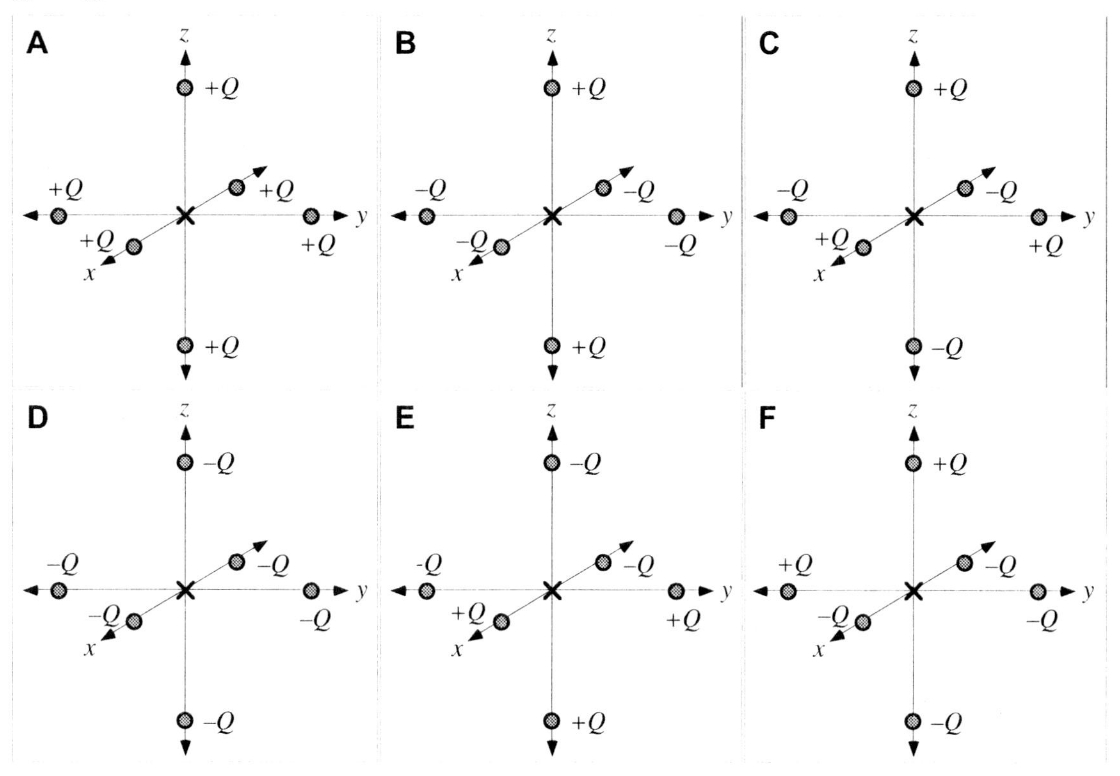

Rank the magnitude of the electric field at the origin.

Greatest 1 _____ 2 _____ 3 _____ 4 _____ 5 _____ 6 _____ Least

OR, the electric field at the origin is the same (but not zero) for all six cases. ____

OR, the electric field at the origin is zero for all six cases. ____

OR, the ranking for the electric field at the origin cannot be determined. ____

Carefully explain your reasoning.

How sure were you of your ranking? (circle one)

Basically Guessed   Sure   Very Sure

1 2 3 4 5 6 7 8 9 10

ET5-RT7: POTENTIAL NEAR CHARGES—ELECTRIC FIELD

In each situation below, electric charges are arranged at equal distances from a point with a specified potential. All of the charges have the same magnitude but the signs of the charges are not given.

A

$V = 0$

B

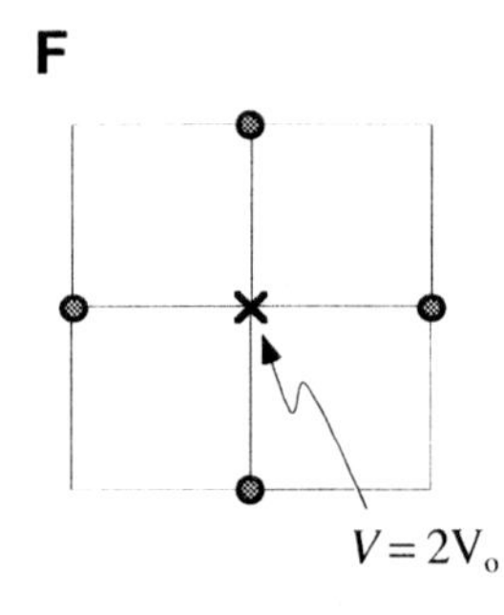

C

$V = 2V_o$

D

$V = -4V_o$

E

$V = -2V_o$

F

$V = 2V_o$

Rank the strength (magnitude) of the electric field at the point where the potential is given.

Greatest 1 _____ 2 _____ 3 _____ 4 _____ 5 _____ 6 _____ Least

OR, the electric field has the same non-zero strength in all these situations. _____

OR, the electric field is zero for these situations. _____

OR, the ranking for the electric field cannot be determined. _____

Carefully explain your reasoning.

How sure were you of your ranking? (circle one)

Basically Guessed					Sure				Very Sure
1	2	3	4	5	6	7	8	9	10

ET5-RT8: THREE CHARGES IN A LINE—ELECTRIC FIELD

Shown below are six situations where three point charges are placed in a row. The magnitudes and signs of the charges are given in the figures.

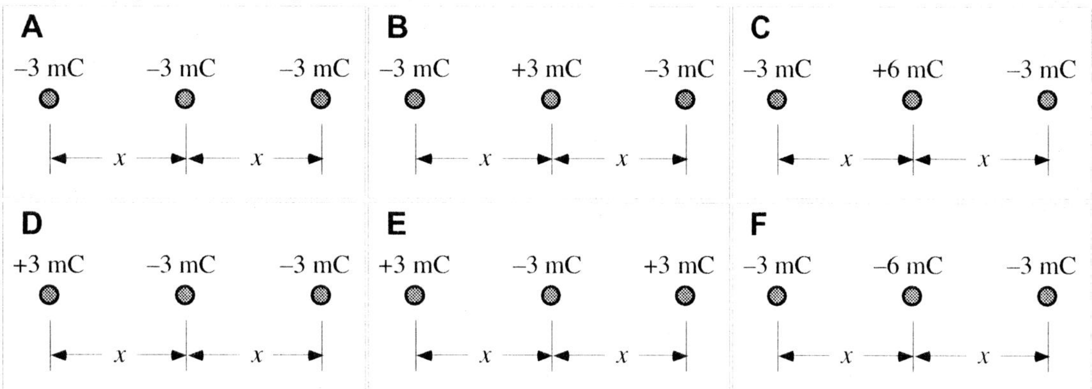

Rank the magnitude of the electric field that is exerting a force on the right-hand charge.

Greatest 1________ 2________ 3________ 4________ 5________ 6________ Least

OR, the electric field is the same but not zero for all six situations. _____

OR, the electric field is zero for all six situations. _____

OR, the ranking for the electric field cannot be determined. _____

Carefully explain your reasoning.

How sure were you of your ranking? (circle one)

Basically Guessed   Sure   Very Sure

1   2   3   4   5   6   7   8   9   10

ET5-RT9: Potential vs Position Graphs I—Electric Field

The graphs below of electric potential versus position in the x-direction are for regions in which there may be electric fields.

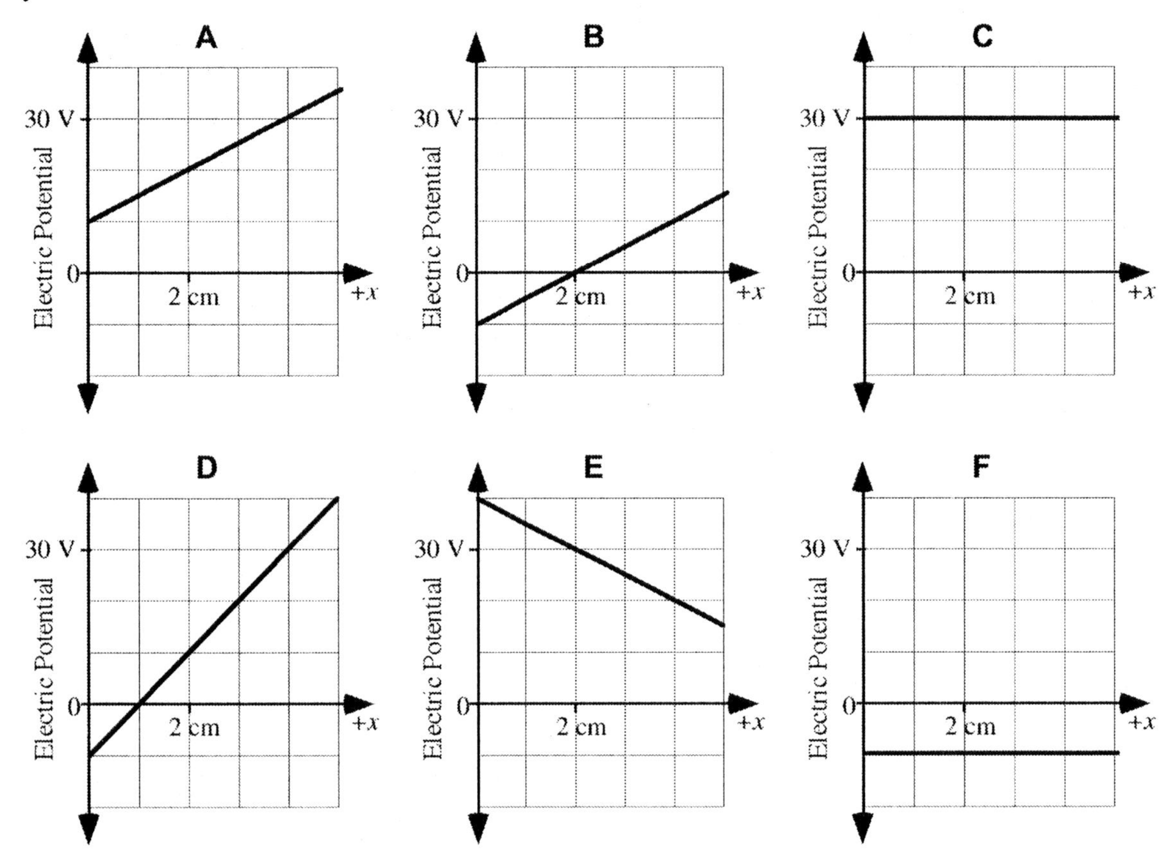

Rank the electric field at $x = 2$ cm in these regions. An electric field in the positive x-direction is greater than an electric field in the negative x-direction.

Greatest positive 1 _______ 2 _______ 3 _______ 4 _______ 5 _______ 6 _______ Greatest negative

OR, the electric field is the same (but not zero) at x = 2 cm for all of these regions. _____

OR, the electric field is zero at x = 2 cm for all of these regions. _____

OR, the ranking for the electric field cannot be determined. _____

Carefully explain your reasoning.

How sure were you of your ranking? (circle one)

Basically Guessed				Sure					Very Sure
1	2	3	4	5	6	7	8	9	10

ET5-RT10: POTENTIAL VS POSITION GRAPH II—ELECTRIC FIELD

The graph of electric potential versus position in the x-direction for a region is graphed below. Six positions in this region are labeled.

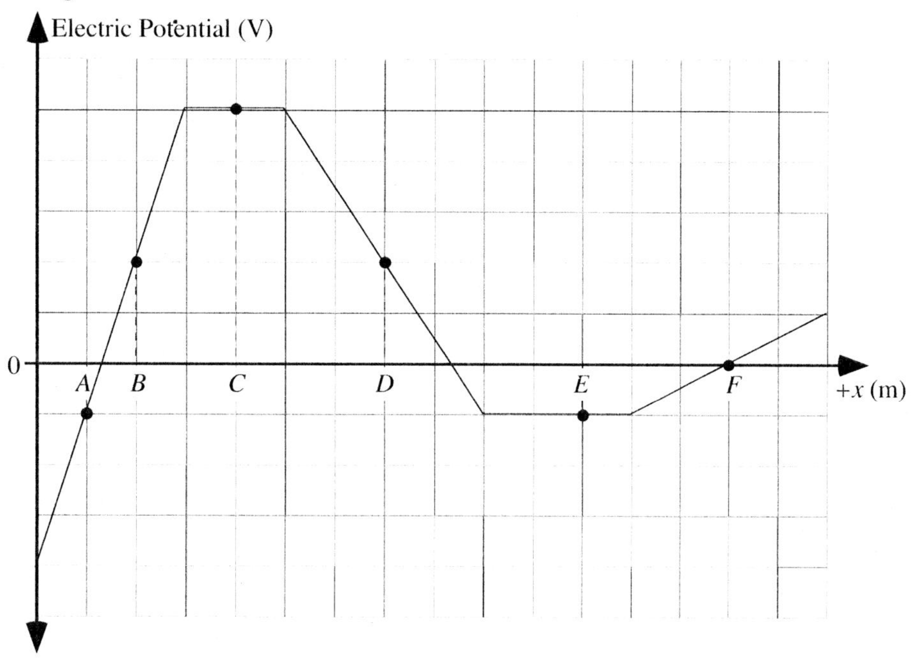

Rank the electric field at the labeled positions. An electric field in the positive x-direction is greater than an electric field in the negative x-direction.

Greatest positive 1 ______ 2 ______ 3 ______ 4 ______ 5 ______ 6 ______ Greatest negative

OR, the electric field is the same (but not zero) for all of these positions. _____

OR, the electric field is zero for all of these positions. _____

OR, the ranking of the electric field cannot be determined at these positions. _____

Carefully explain your reasoning.

How sure were you of your ranking? (circle one)

Basically Guessed				Sure					Very Sure
1	2	3	4	5	6	7	8	9	10

ET5-RT11: POTENTIAL VS POSITION GRAPH III—ELECTRIC FIELD

The graph of electric potential versus position in the *x*-direction for a region is graphed below. Six positions in this region are labeled.

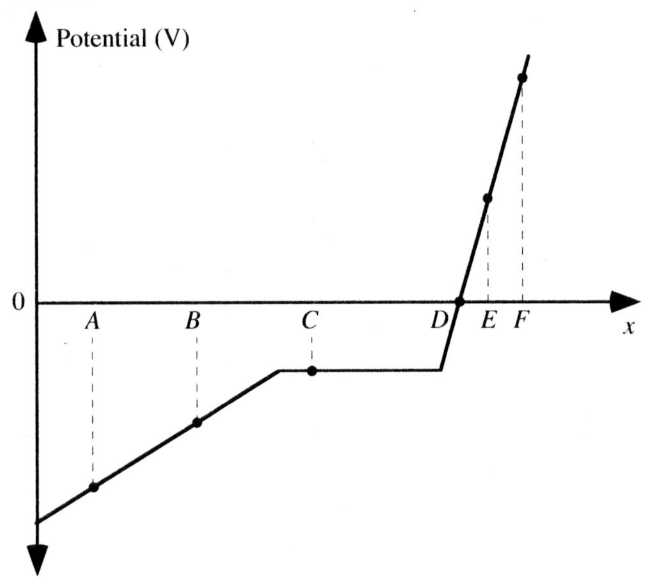

Rank the strength (magnitude) of the electric field in the *x*-direction at the labeled positions.

Greatest 1 ______ 2 ______ 3 ______ 4 ______ 5 ______ 6 ______ Least

OR, the electric field is the same (but not zero) for all six positions. _____

OR, the electric field is zero for all six positions. _____

OR, the ranking of the electric field cannot be determined at these positions. ______

Carefully explain your reasoning.

How sure were you of your ranking? (circle one)

Basically Guessed					Sure				Very Sure
1	2	3	4	5	6	7	8	9	10

ET5-RT12: POINT CHARGES IN TWO DIMENSIONS—ELECTRIC FIELD

In each situation shown below, charged particles are fixed on grids having the same spacing. Each charge Q on this page has the same magnitude with signs given in the diagrams.

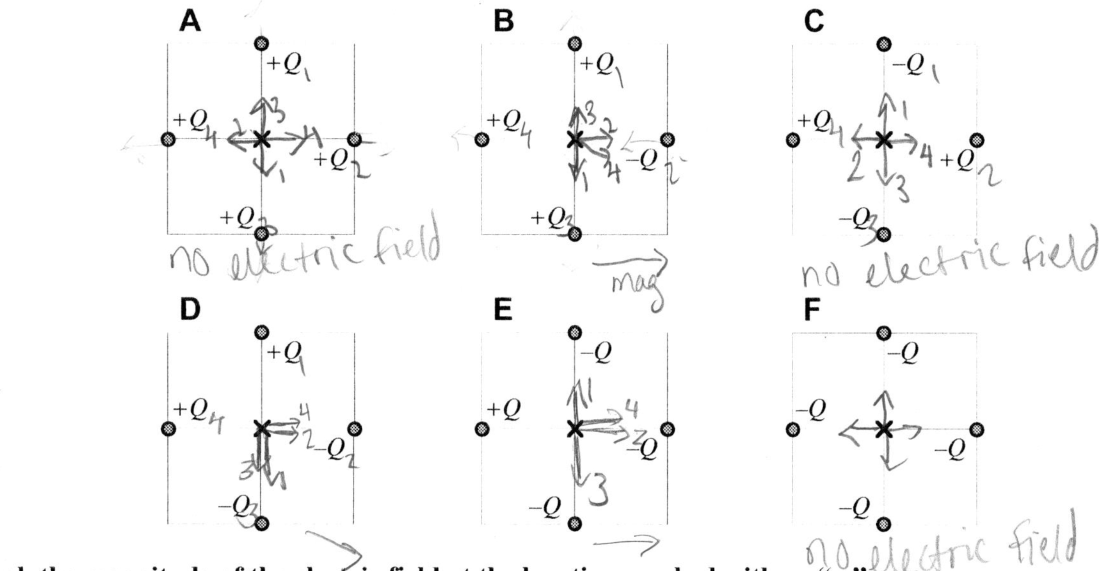

Rank the magnitude of the electric field at the location marked with an "x."

Greatest 1 D 2 B 3 E 4 A 5 C 6 F Least

OR, the electric field is the same (but not zero) for all six cases. ____

OR, the electric field is zero for all six cases. ____

OR, the ranking for the electric field cannot be determined. ____

D > B = E > A = C = F

Carefully explain your reasoning.

How sure were you of your ranking? (circle one)

Basically Guessed				Sure				Very Sure
1	2	3	4	5	6	7	8	(9)	10

ET5-RT13: ELECTRIC FIELD LINES—ELECTRIC FIELD

Shown are electric field lines that represent the electric field in a region of space. The charges that produced the electric field are not shown.

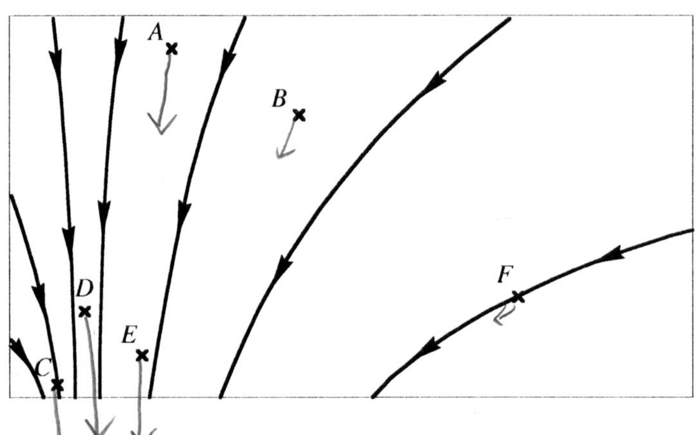

Rank the magnitude of the electric field for the points labeled *A* through *F*.

Greatest 1 _____ 2 _____ 3 _____ 4 _____ 5 _____ 6 _____ Least

OR, the electric field is the same (but not zero) for all six locations. _____

OR, the electric field is zero for all six locations. _____

OR, the ranking of the electric field cannot be determined. _____

Explain how you determined your ranking.

How sure were you of your ranking? (circle one)

Basically Guessed — Sure — Very Sure

1 2 3 4 5 6 7 8 9 10

ET5-RT14: Charged Curved Rod—Electric Field

Point P is located at the center of curvature of a curved, charged, insulating rod as shown at left below. For each of the five cases A-E, the charge density on the rod varies according to the graphs shown below but the total charge is the same.

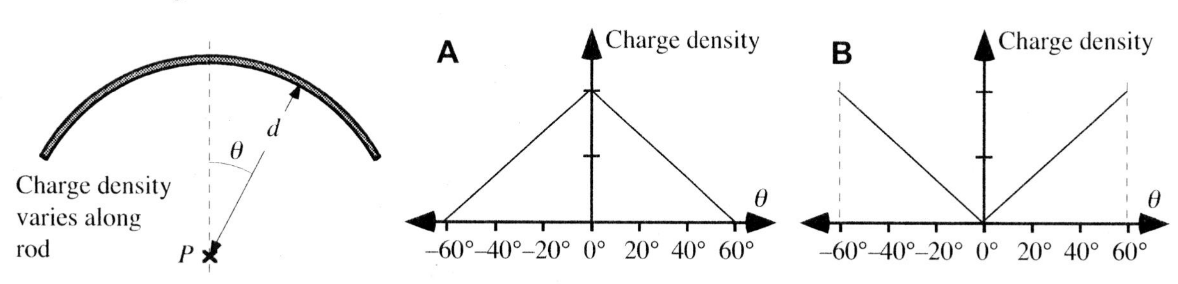

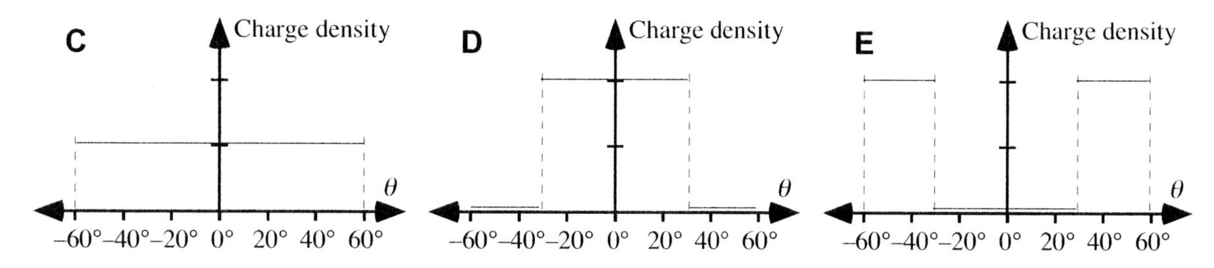

Rank the magnitude of the electric field at *P* due to the charge in the curved rod.

Greatest 1 ________ 2 ________ 3 ________ 4 ________ 5 ________ Least

OR, the electric field at P is the same (but not zero) for all five cases. ____

OR, the electric field at P is zero for all five cases. ____

OR, the ranking for the electric field at P cannot be determined. ____

Carefully explain your reasoning.

How sure were you of your ranking? (circle one)

Basically Guessed				Sure					Very Sure
1	2	3	4	5	6	7	8	9	10

ET5-RT15: Three-Dimensional Locations Within a Uniform Electric Field—Field

All the labeled points are within a region of space with a uniform electric field. The electric field points toward the top of the page (that is, in the positive z-direction).

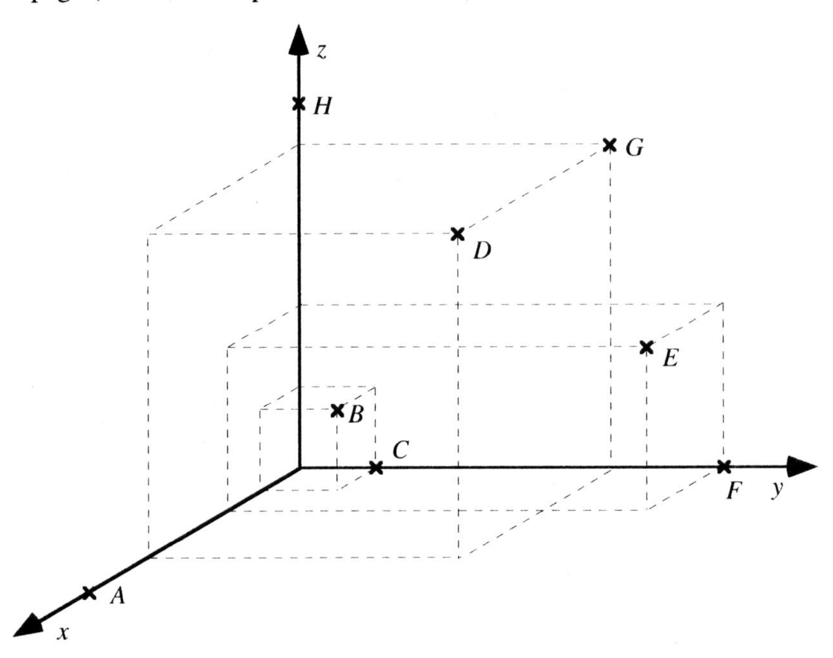

Rank the strength (magnitude) of the electric field at the labeled points.

Greatest 1 ______ 2 ______ 3 ______ 4 ______ 5 ______ 6 ______ 7 ______ 8 ______ Least

OR, the strength of the electric field is the same (but not zero) for all of these points. ____

OR, the strength of the electric field is zero for all these points. ____

OR, the ranking for the electric field strength cannot be determined for all of these points. ____

Carefully explain your reasoning.

How sure were you of your ranking? (circle one)

Basically Guessed				Sure					Very Sure
1	2	3	4	5	6	7	8	9	10

ET5-RT16: POINT CHARGE INSIDE AN INSULATING SHELL—ELECTRIC FIELD

A point charge is enclosed by an uncharged plastic (insulating) shell, but is offset from the center of the shell. Four points are labeled in the diagram below. Points B and C are the same distance from the center of the shell, and points A and D are the same distance from the center of the shell.

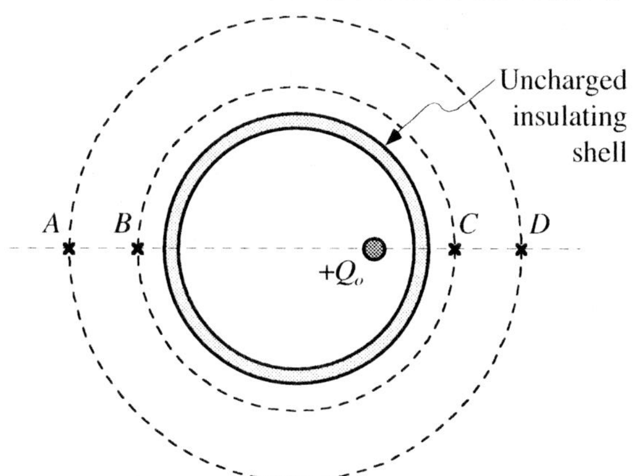

Rank the strength (magnitude) of the electric field at the labeled points.

Greatest 1 _______ 2 _______ 3 _______ 4 _______ Least

OR, the magnitude of the electric field is nonzero but is the same for all four points. ____

OR, the electric field is zero at all four points. ____

OR, the ranking for the magnitude of the electric field cannot be determined. ____

Carefully explain your reasoning.

How sure were you of your ranking? (circle one)

Basically Guessed — Sure — Very Sure

1 2 3 4 5 6 7 8 9 10

ET5-RT17: POINT CHARGE INSIDE A CONDUCTING SHELL—ELECTRIC FIELD

A point charge is enclosed by an uncharged metal (conducting) shell, but is offset from the center of the shell. Four points are labeled in the diagram below. Points *B* and *C* are the same distance from the center of the shell, and points *A* and *D* are the same distance from the center of the shell.

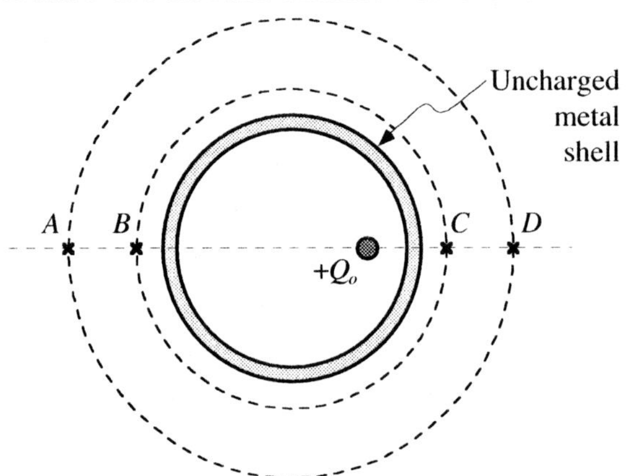

Rank the magnitude of the electric field at the labeled points.

Greatest 1 _______ 2 _______ 3 _______ 4 _______ Least

OR, the magnitude of the electric field is nonzero but is the same for all four points. _____

OR, the electric field is zero at all four points. _____

OR, the ranking for the magnitude of the electric field cannot be determined. _____

Carefully explain your reasoning.

How sure were you of your ranking? (circle one)

Basically Guessed — Sure — Very Sure

1 2 3 4 5 6 7 8 9 10

ET5-RT18: EQUIPOTENTIAL SURFACES—ELECTRIC FIELD

The diagrams below show portions of four different regions with electric fields. The dashed lines represent cross-sections of flat electric equipotential surfaces or sheets. The potentials for several sheets in each region are labeled. The distance from the leftmost equipotential surface and the rightmost one is 20 cm in each case.

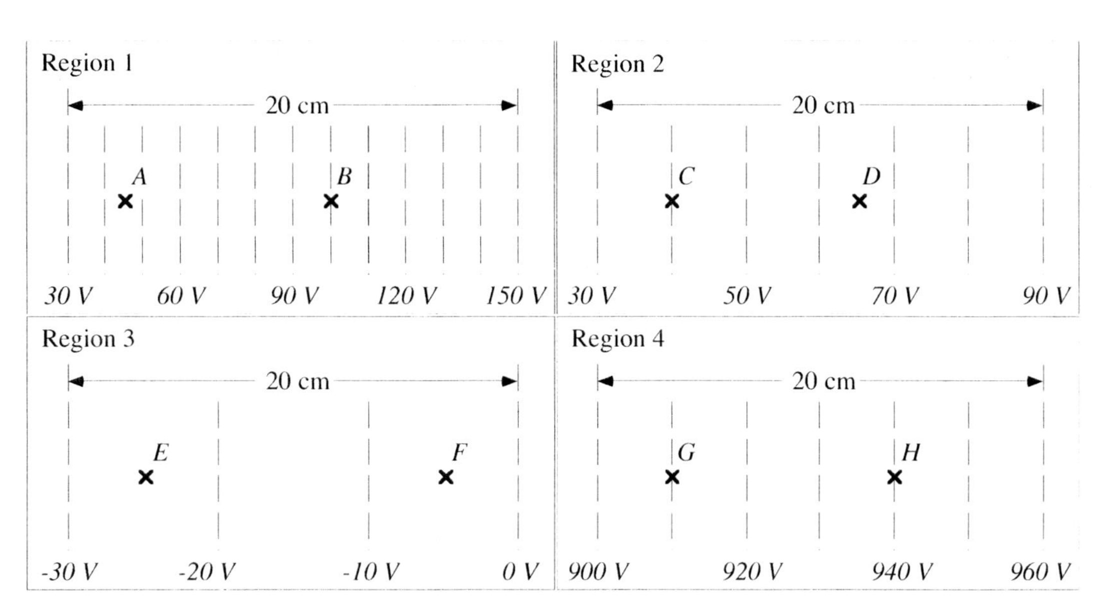

Rank the magnitude of the electric field at the labeled points.

Greatest 1 ______ 2 ______ 3 ______ 4 ______ 5 ______ 6 ______ 7 ______ 8 ______ Least

OR, the electric field is the same but not zero for all of these points. _______

OR, the electric field is zero in all of these points. _______

OR, the electric field cannot be determined for some or all of these points. _______

Carefully explain your reasoning.

How sure were you of your ranking? (circle one)

Basically Guessed — Sure — Very Sure

1 2 3 4 5 6 7 8 9 10

ET6-RT1: THREE-DIMENSIONAL LOCATIONS IN A CONSTANT ELECTRIC POTENTIAL—WORK

The electric potential is a constant 12 volts everywhere in a three-dimensional region. Within that region are points located at the corners of two cubes as shown below. The small cube has edges of 1 centimeter length, and the larger cube has edges of 3 centimeters length. A +2 μC charge is placed at the origin.

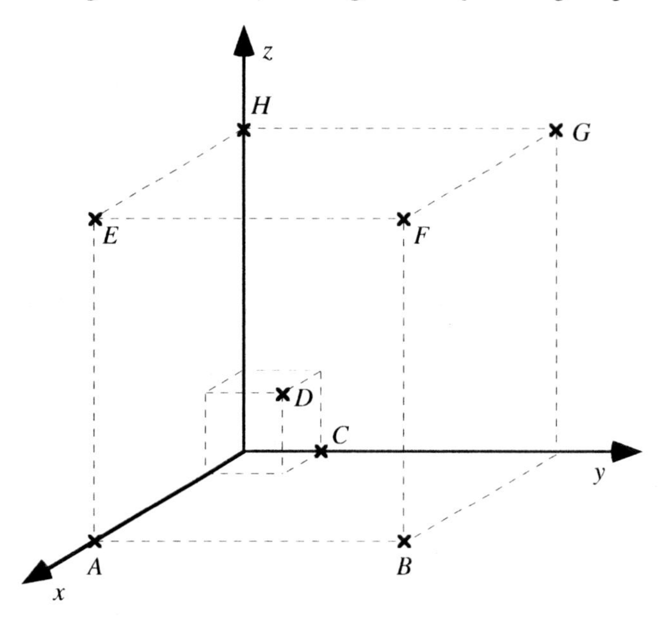

Rank the work required to move the +2 μC charge along a straight-line path from the origin to each of the labeled points.

Greatest 1 ______ 2 ______ 3 ______ 4 ______ 5 ______ 6 ______ 7 ______ 8 ______ Least

OR, the work required is the same but not zero for moving to all of these points. ____

OR, the work required is zero for moving to all of these points. ____

OR, the ranking for the work required cannot be determined for all of these points. ____

Carefully explain your reasoning.

How sure were you of your ranking? (circle one)

Basically Guessed				Sure					Very Sure
1	2	3	4	5	6	7	8	9	10

ET6-RT2: Three Charge System—Electric Potential Energy

In each case shown below, charged particles are arranged in identical equilateral triangles. The particles are all the same size and mass but they have different charges.

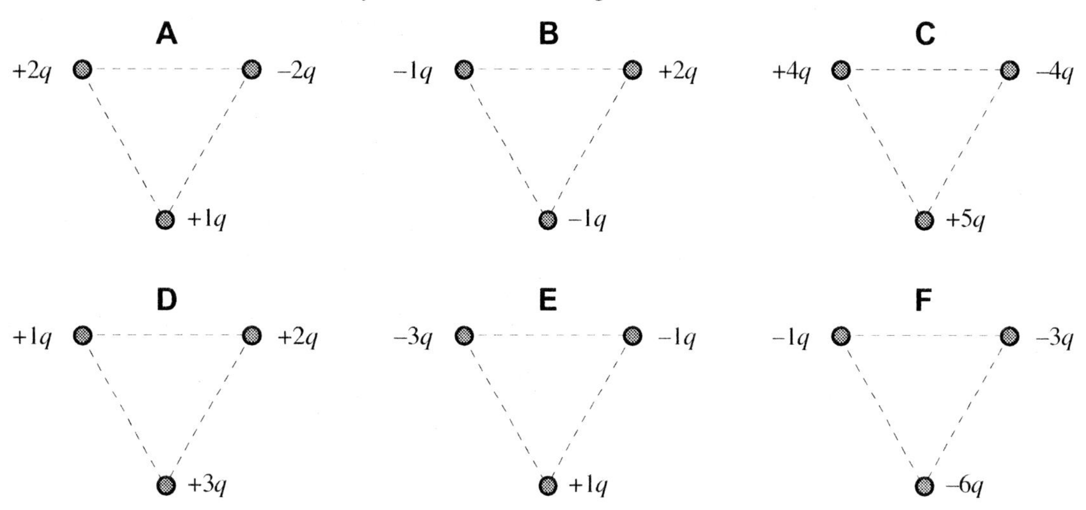

Rank the electric potential energy of the six arrangements.

Greatest 1 _______ 2 _______ 3 _______ 4 _______ 5 _______ 6 _______ Least

OR, the electric potential energy is the same but not zero for every case. ______

OR, the electric potential energy is zero for every case. ______

OR, the ranking of the electric potential energy cannot be determined. ______

Carefully explain your reasoning.

How sure were you of your ranking? (circle one)

Basically Guessed   Sure   Very Sure

1   2   3   4   5   6   7   8   9   10

ET6-RT3: ELECTRON IN EQUIPOTENTIAL SURFACES—KINETIC ENERGY CHANGE

The diagrams below show regions of space that contain electric fields. The dotted lines represent cross-sections of electric equipotential surfaces or sheets. The distance from the leftmost equipotential to the rightmost one is 20 cm in each case. An electron is placed at point *S* with an initial velocity of 100 m/s to the right.

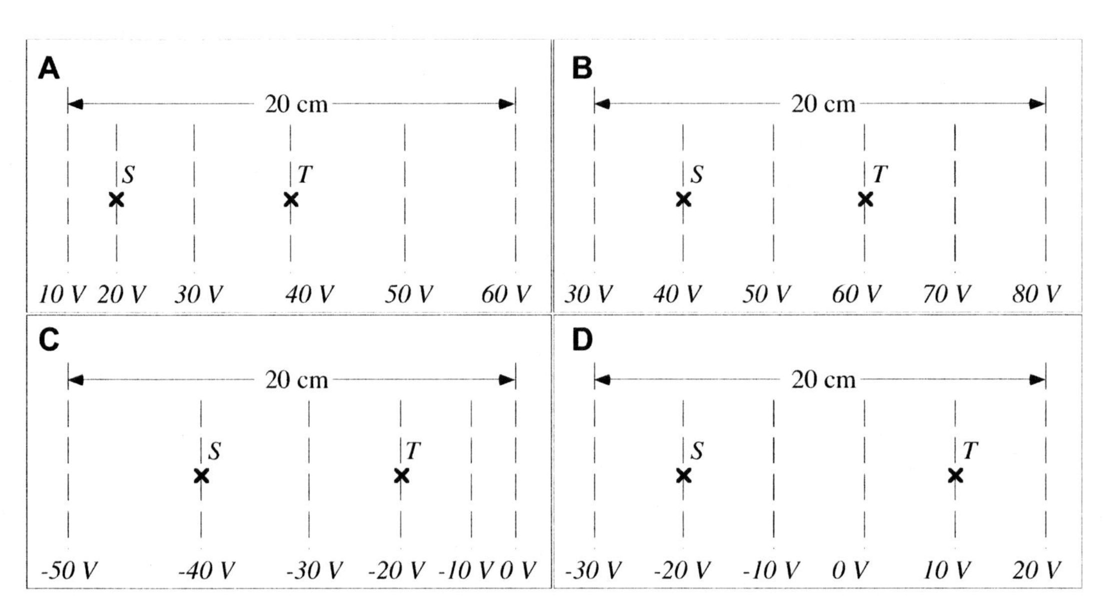

Rank the change in kinetic energy of the electron as it travels between points *S* and *T* in each case.

Greatest 1 _______ 2 _______ 3 _______ 4 _______ Least

OR, the change in kinetic energy is the same (but not zero) for all of these cases. _______

OR, the change in kinetic energy is zero in all of these cases. _______

OR, the ranking for the change in kinetic energy cannot be determined. _______

Carefully explain your reasoning.

How sure were you of your ranking? (circle one)

Basically Guessed — Sure — Very Sure

1 2 3 4 5 6 7 8 9 10

ET6-RT4: CHARGES AND EQUIPOTENTIALS—WORK

The dashed lines in the figure below represent equipotentials (with magnitudes labeled in the margins of the drawing) in a region in which there is an electric field. Also shown are four circles at different distances from point *P*. Six points on these circles are labeled *A* to *F*. A positive point charge at rest is moved in a straight line from *P* to each of the six labeled points in turn. It also ends at rest at these points.

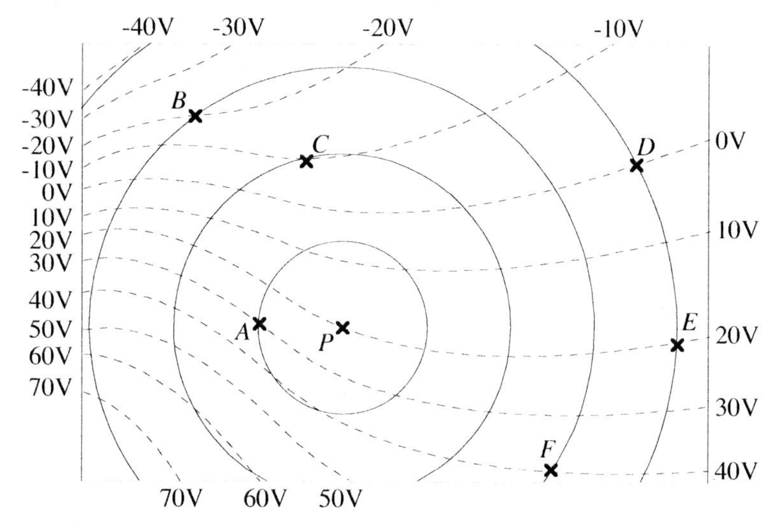

Rank the work that must be done by an external agent in order to move the positive charge from *P* to the labeled points *A-F*.

Greatest 1 _______ 2 _______ 3 _______ 4 _______ 5 _______ 6 _______ Least

OR, the same (nonzero) amount of work is required to move to all six points. _____

OR, no work is required to move to any of the six labeled points. _____

OR, we cannot determine the ranking from the given information. _____

Carefully explain your reasoning.

How sure were you of your ranking? (circle one)

Basically Guessed — Sure — Very Sure

1 2 3 4 5 6 7 8 9 10

ET8-RT1: FOUR CHARGES IN TWO DIMENSIONS—ELECTRIC POTENTIAL

In each situation shown below, small charged particles are fixed on grids having the same spacing. Each charge Q on this page has the same magnitude with the signs indicated in the diagrams.

A

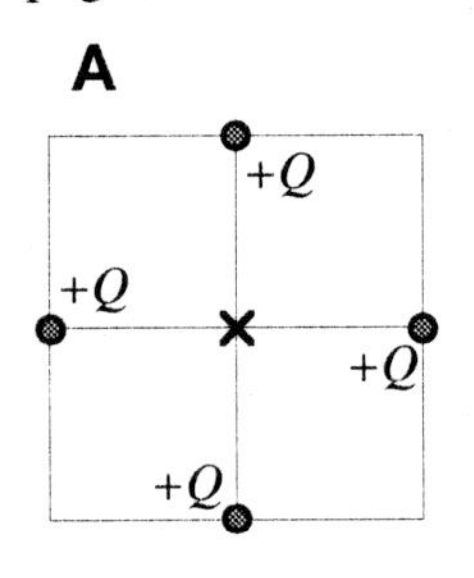

B

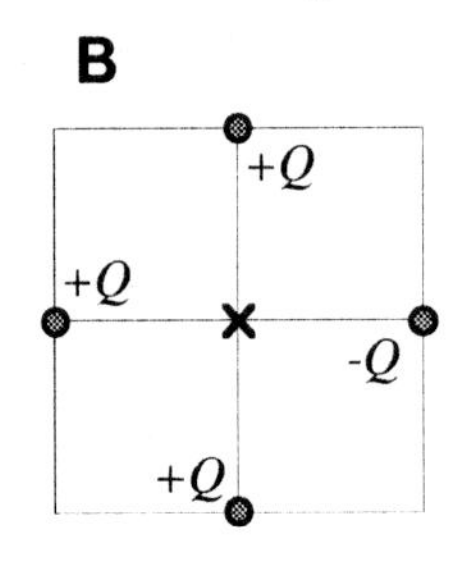

C

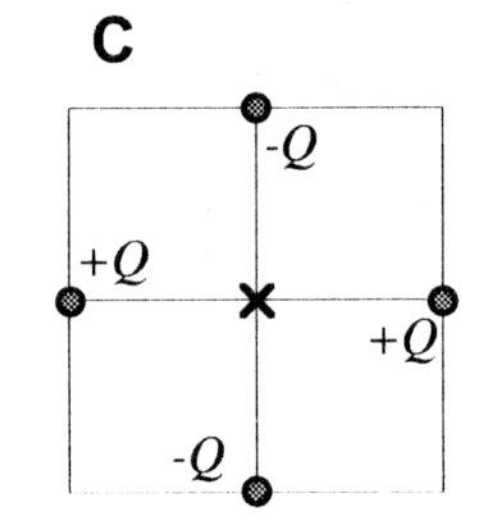

D

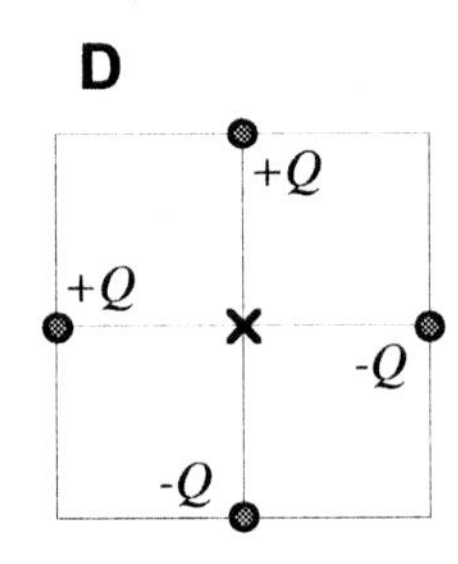

E

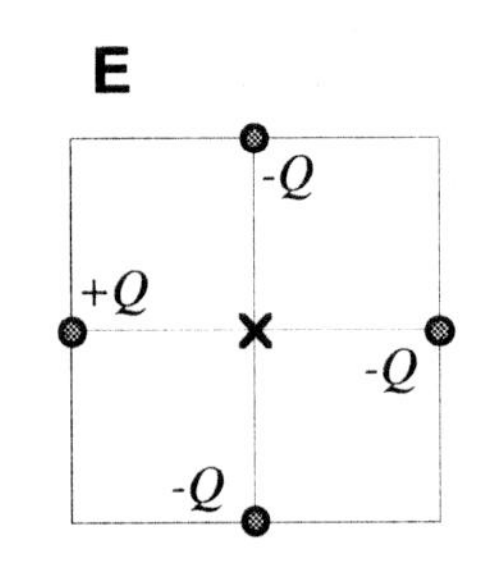

F

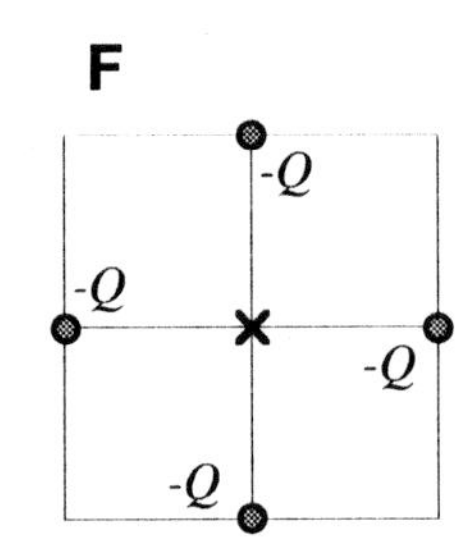

Rank the electric potential at the location marked with an "x."

Greatest positive 1 _______ 2 _______ 3 _______ 4 _______ 5 _______ 6 _______ Greatest negative

OR, the electric potential at the center is the same (but not zero) for all six cases. ____

OR, the electric potential at the center is zero for all six cases. ____

OR, the ranking for the electric potential at the center cannot be determined. ____

Carefully explain your reasoning.

How sure were you of your ranking? (circle one)

Basically Guessed				Sure					Very Sure
1	2	3	4	5	6	7	8	9	10

ET8-RT2: Points near a Pair of Equal Opposite Charges—Potential

Two equal and opposite charges are fixed in space at the locations shown. Seven points in the vicinity of these charges are labeled *A* - *G*.

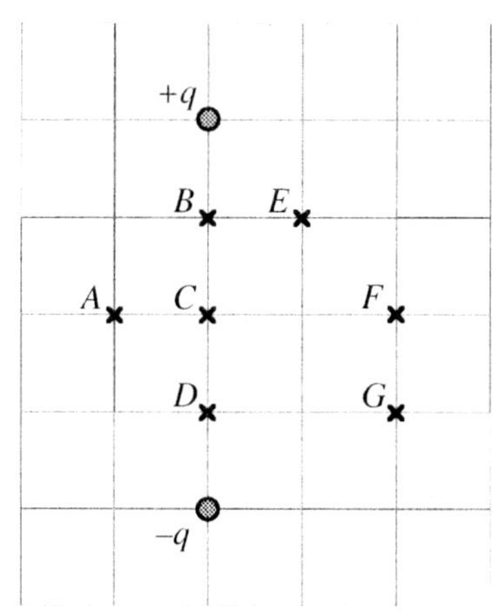

Rank the electric potential at the labeled points.

Greatest 1 _______ 2 _______ 3 _______ 4 _______ 5 _______ 6 _______ 7 _______ Least

OR, the electric potential is the same (but not zero) at all seven points. ____

OR, the electric potential is zero at all seven points. ____

OR, the ranking for the electric potential cannot be determined. ____

Carefully explain your reasoning.

How sure were you of your ranking? (circle one)

Basically Guessed Sure Very Sure

1 2 3 4 5 6 7 8 9 10

ET8-RT3: PAIRS OF CHARGED CONNECTED CONDUCTORS—ELECTRIC POTENTIAL

Three pairs of charged conducting spheres connected with wires and switches. The spheres are very far apart. The large spheres have twice the radius of the small spheres. Each sphere on the left has a charge of +20 nC and each sphere on the right has a charge of +70 nC before the switches are closed.

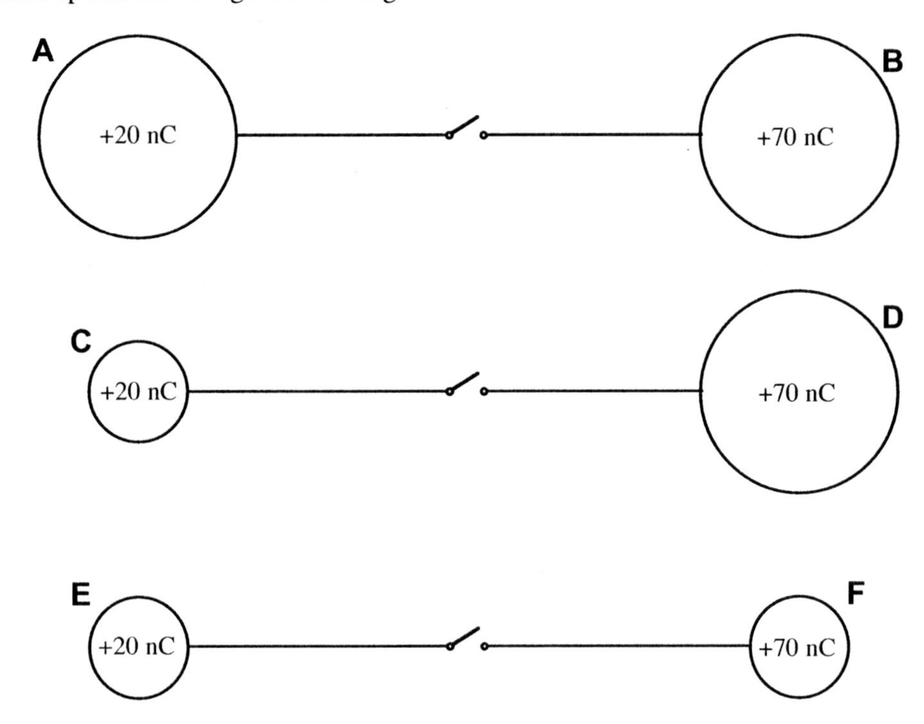

Rank the electric potential on the spheres after all of the switches are closed.

Greatest 1 _______ 2 _______ 3 _______ 4 _______ 5 _______ 6 _______ Least

OR, the electric potential is the same but not zero for all spheres. _____

OR, the electric potential is zero for all spheres. _____

OR, the ranking for the electric potential cannot be determined. _____

Carefully explain your reasoning.

How sure were you of your ranking? (circle one)

Basically Guessed					Sure				Very Sure
1	2	3	4	5	6	7	8	9	10

ET8-RT4: Charged Curved Rod—Electric Potential

A point P is located at the center of curvature of a curved, charged, insulating rod as shown at left below. For each of the five cases A-E, the charge density on the rod varies as shown on the graphs below, but the total charge is the same.

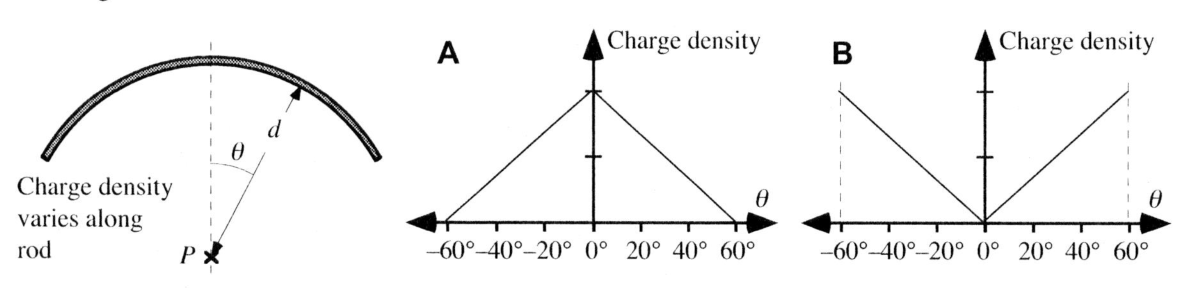

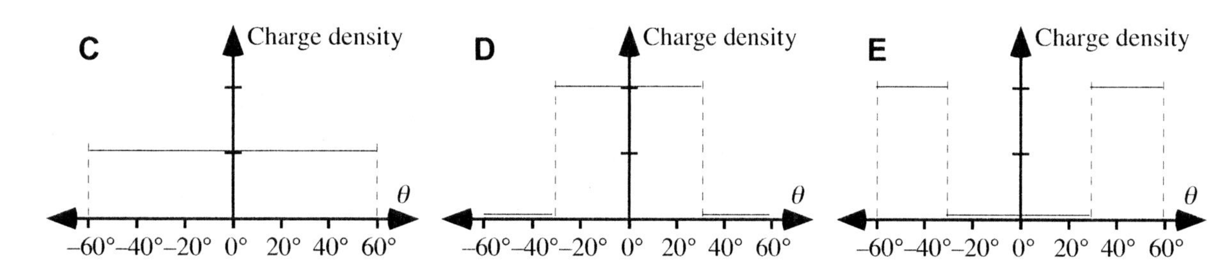

Rank the electric potential at *P* due to the charge in the curved rod for the five cases.

Greatest 1 _______ 2 _______ 3 _______ 4 _______ 5 _______ Least

OR, the electric potential at P is the same (but not zero) for all five cases. ____

OR, the electric potential at P is zero for all five cases. ____

OR, the ranking for the electric potential at P cannot be determined. ____

Carefully explain your reasoning.

How sure were you of your ranking? (circle one)

Basically Guessed — Sure — Very Sure

1 2 3 4 5 6 7 8 9 10

ET8-RT5: TWO LARGE CHARGED PARALLEL SHEETS—POTENTIAL DIFFERENCE

Each diagram below shows two very large parallel insulating sheets. (Only a small portion near the center of the sheets is shown; the distance between the sheets is very small compared to the dimensions of the sheet.) The sheets are uniformly charged with charge densities and separations as shown.

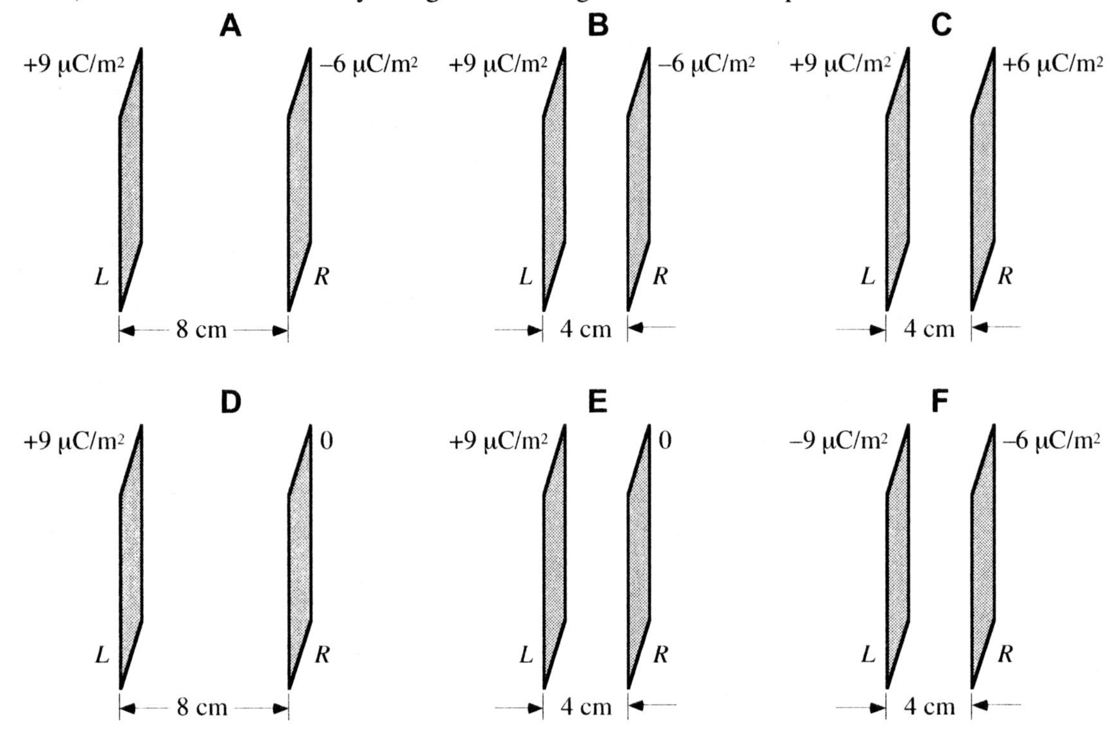

Rank the absolute value of the electric potential difference ($V_L - V_R$) between the sheets.

Greatest 1 _______ 2 _______ 3 _______ 4 _______ 5 _______ 6 _______ Least

OR, the electric potential difference is the same for all six situations. ____

OR, the ranking for the electric potential difference cannot be determined. ____

Carefully explain your reasoning.

How sure were you of your ranking? (circle one)

Basically Guessed — Sure — Very Sure

1 2 3 4 5 6 7 8 9 10

ET8-RT6: THREE-DIMENSIONAL LOCATIONS NEAR A POINT CHARGE—ELECTRIC POTENTIAL

There is a positive point charge $+q$ located at (0, 3, 0) as shown in the three-dimensional region below. Within that region are points located on the corners of two cubes as shown below. The small cube has edges of 1 centimeter length, and the larger cube has edges of 3 centimeter length.

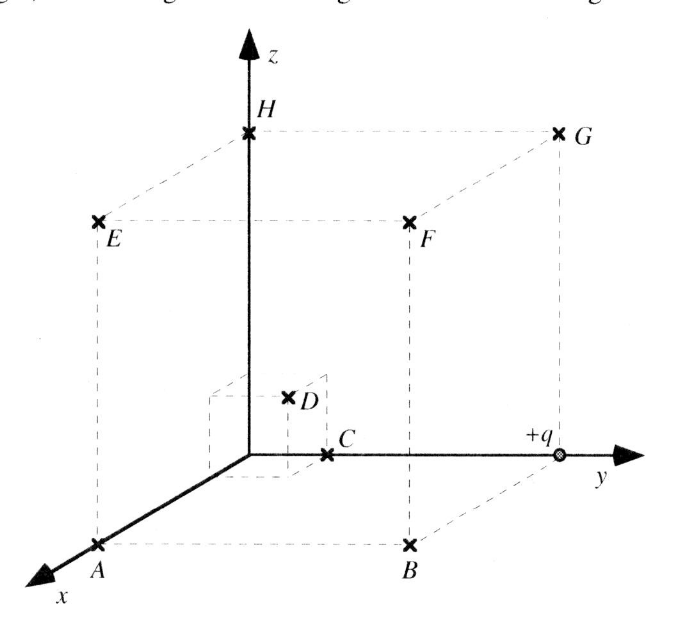

Rank the electric potential at the labeled points.

Greatest 1 _______ 2 _______ 3 _______ 4 _______ 5 _______ 6 _______ 7 _______ 8 _______ Least

OR, the electric potential is the same (but not zero) for all these points. _____

OR, the electric potential is zero for all these points. _____

OR, the ranking for the electric potential cannot be determined for all these points. _____

Carefully explain your reasoning.

How sure were you of your ranking? (circle one)

Basically Guessed				Sure					Very Sure
1	2	3	4	5	6	7	8	9	10

ET8-RT7: THREE-DIMENSIONAL LOCATIONS IN A UNIFORM ELECTRIC FIELD—POTENTIAL

All the points shown below are within a region of space with a uniform electric field. The electric field points toward the top of the page (that is, in the positive z-direction).

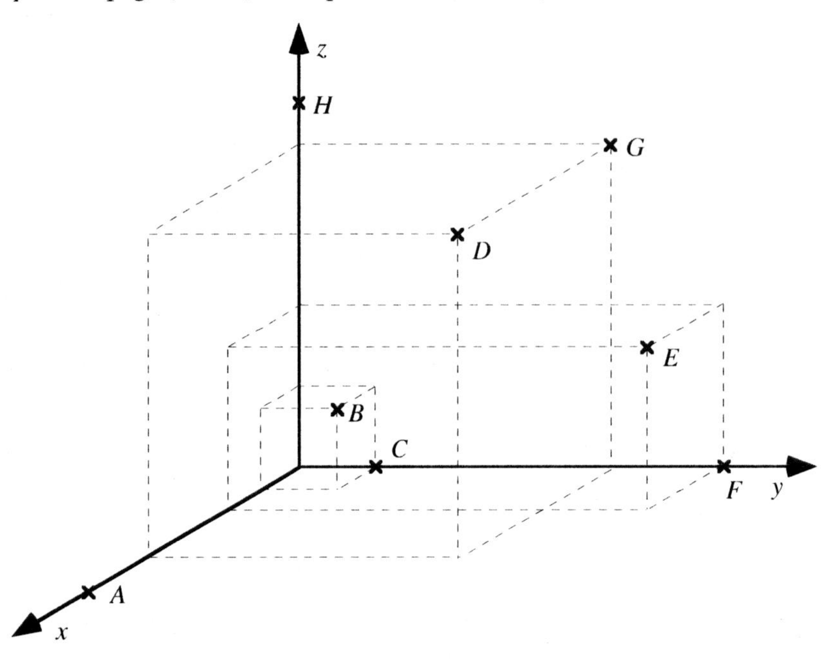

Rank the electric potential at the labeled points.

Greatest 1 _______ 2 _______ 3 _______ 4 _______ 5 _______ 6 _______ 7 _______ 8 _______ Least

OR, the electric potential is the same (but not zero) for all of these points. _____

OR, the electric potential is zero for all of these points. _____

OR, the ranking for the electric potential cannot be determined for all of these points. _____

Carefully explain your reasoning.

How sure were you of your ranking? (circle one)

Basically Guessed				Sure					Very Sure
1	2	3	4	5	6	7	8	9	10

ET8-RT8: SIX CHARGES IN THREE DIMENSIONS—ELECTRIC POTENTIAL

Six point charges are all the same distance away from the origin as shown in these six diagrams. All charges are either $+Q$ or $-Q$.

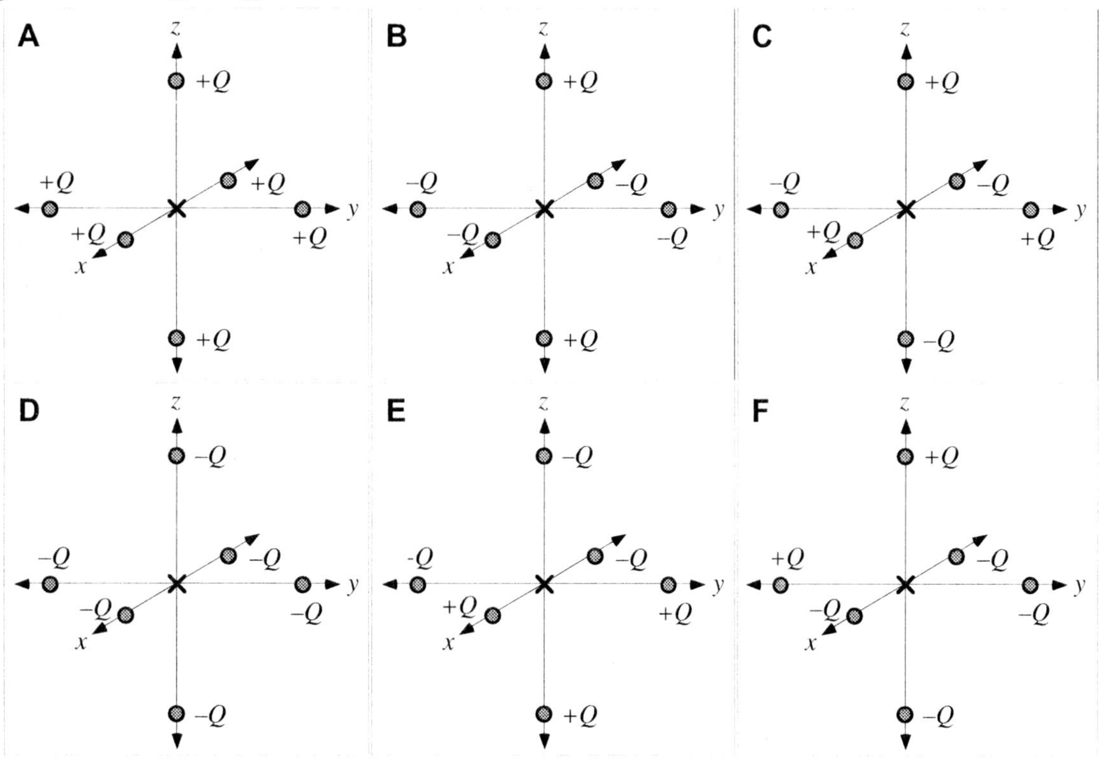

Rank the electric potential at the origin in each of the six cases above.

Greatest positive 1 _____ 2 _____ 3 _____ 4 _____ 5 _____ 6 _____ Greatest negative

OR, the electric potential at the origin is the same (but not zero) for all six cases. ____

OR, the electric potential at the origin is zero for all six cases. ____

OR, the ranking for the electric potential at the origin cannot be determined. ____

Briefly explain how you determined your ranking.

How sure were you of your ranking? (circle one)

Basically Guessed					Sure				Very Sure
1	2	3	4	5	6	7	8	9	10

ET8-RT9: SPHERICAL CONDUCTING SHELL—ELECTRIC POTENTIAL

A spherical conducting shell with a radius of 200 cm has a charge of +300 μC.

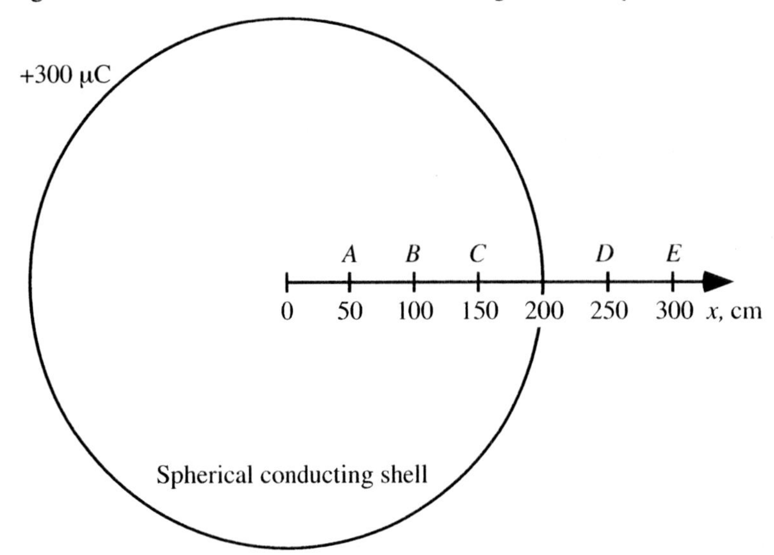

Rank the electric potential at the labeled points.

Greatest 1_________ 2_________ 3_________ 4_________ 5_________ Least

OR, the electric potential is the same but not zero for all five points. ____

OR, the electric potential is zero for all five points. ____

OR, the ranking for the electric potential cannot be determined. ____

Carefully explain your reasoning.

How sure were you of your ranking? (circle one)

Basically Guessed — Sure — Very Sure

1 2 3 4 5 6 7 8 9 10

eT8-RT10: Systems of Eight Point Charges—Potential

Shown below are three arrangements of eight positive point charges fixed on grids having the same spacing. The point charges all have the same charge and mass.

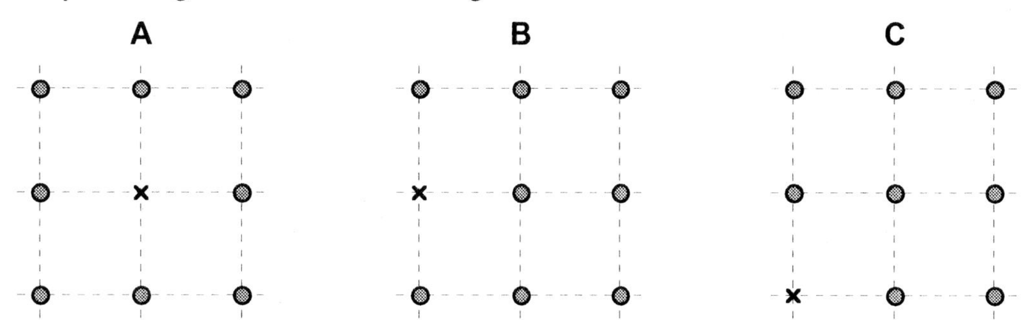

Rank the electric potential at the point marked by an "x" in these arrangements.

Greatest 1________ 2 ________ 3 ________ Least

OR, the electric potential is the same at these points for all of the arrangements. _____

OR, it is not possible to rank the electric potential at these points. _____

Carefully explain your reasoning.

How sure were you of your ranking? (circle one)

Basically Guessed					Sure				Very Sure
1	2	3	4	5	6	7	8	9	10

ET9-RT1: POINT CHARGES—ELECTRIC FLUX

Each figure below shows a cross-section of a spherical Gaussian surface surrounding a point charge.

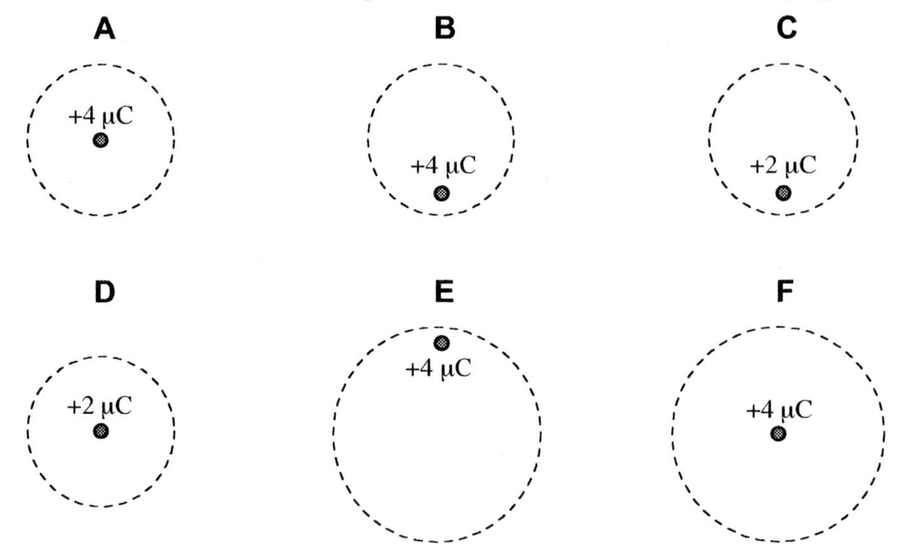

Rank the total electric flux through the given surfaces.

Greatest 1 _______ 2 _______ 3 _______ 4 _______ 5 _______ 6 _______ Least

OR, all of the surfaces have the same (but not zero) electric flux. _____

OR, all of the surfaces have zero electric flux. _____

OR, the ranking of the electric flux cannot be determined. _____

Carefully explain your reasoning.

How sure were you of your ranking? (circle one)

Basically Guessed					Sure				Very Sure
1	2	3	4	5	6	7	8	9	10

ET9-RT2: CHARGED INSULATOR AND CONDUCTOR—ELECTRIC FLUX

The figures below show cross-sections of six spherical Gaussian surfaces labeled A-F. The surfaces on the left enclose a uniformly charged insulator, and the surfaces on the right enclose a charged conductor. The insulator and the conductor have the same total charge $+Q$.

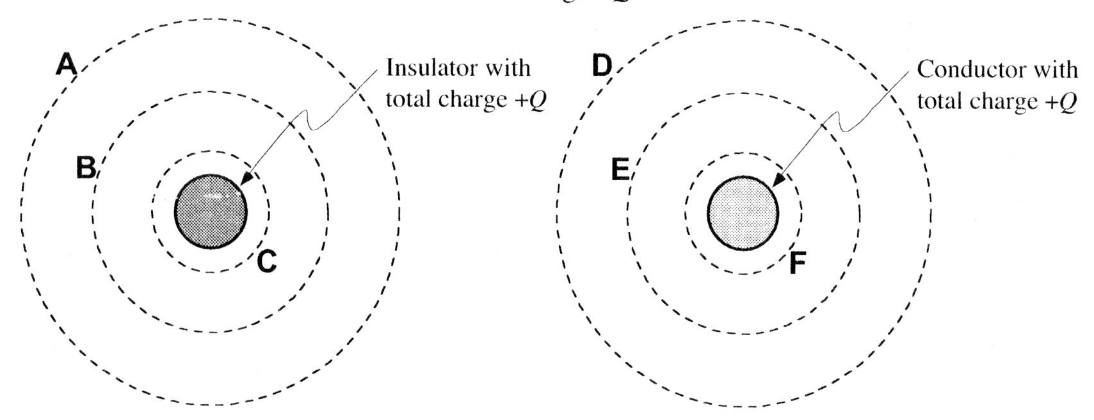

Rank the electric flux through the labeled surfaces.

Greatest 1 _______ 2 _______ 3 _______ 4 _______ 5 _______ 6 _______ Least

OR, all of the surfaces have the same (but not zero) electric flux. _____

OR, all of the surfaces have zero electric flux. _____

OR, the ranking of the electric flux cannot be determined. _____

Carefully explain your reasoning.

How sure were you of your ranking? (circle one)

Basically Guessed				Sure					Very Sure
1	2	3	4	5	6	7	8	9	10

ET9-RT3: INSULATOR AND CONDUCTOR—ELECTRIC FLUX

The figures below show cross-sections of spherical or cubical Gaussian surfaces that intersect solid charged spheres. The total charge $+Q$ is the same for the conducting sphere on the left and the uniformly charged insulating sphere on the right. The radii of surfaces A, C, and D are half that of the conducting and insulating spheres.

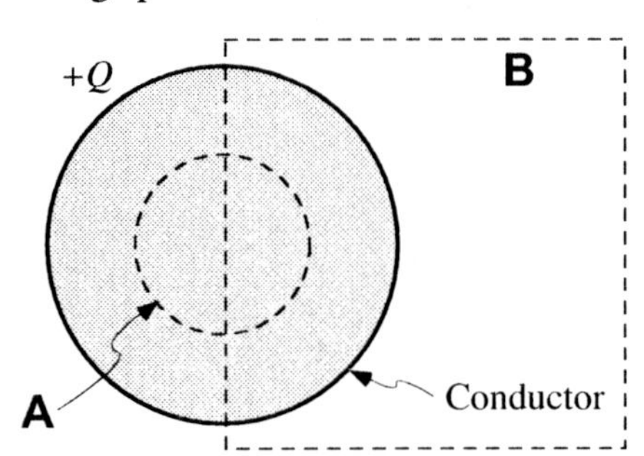

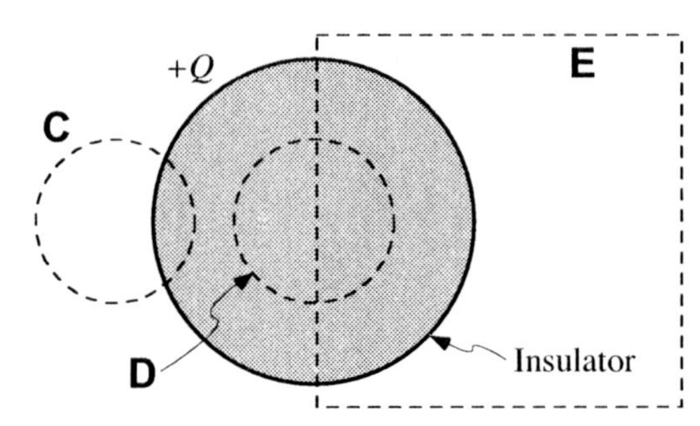

Rank the total electric flux through the surfaces labeled A-E.

Greatest 1 _______ 2 _______ 3 _______ 4 _______ 5 _______ Least

OR, all five surfaces have the same (nonzero) electric flux. _____

OR, all five surfaces have zero electric flux. _____

Or, the ranking of the electric flux cannot be determined with the information provided. _____

Carefully explain your reasoning.

How sure were you of your ranking? (circle one)

Basically Guessed				Sure					Very Sure
1	2	3	4	5	6	7	8	9	10

ET9-RT4: GAUSSIAN CUBES IN NON-UNIFORM ELECTRIC FIELDS—ELECTRIC FLUX

Shown below are six situations where cubical Gaussian surfaces are located within non-uniform electric fields. The fields are produced solely by charge distributions located outside the cubes. The side lengths of the cubes and the total electric flux through five of the six faces of the cubes are given in the figures.

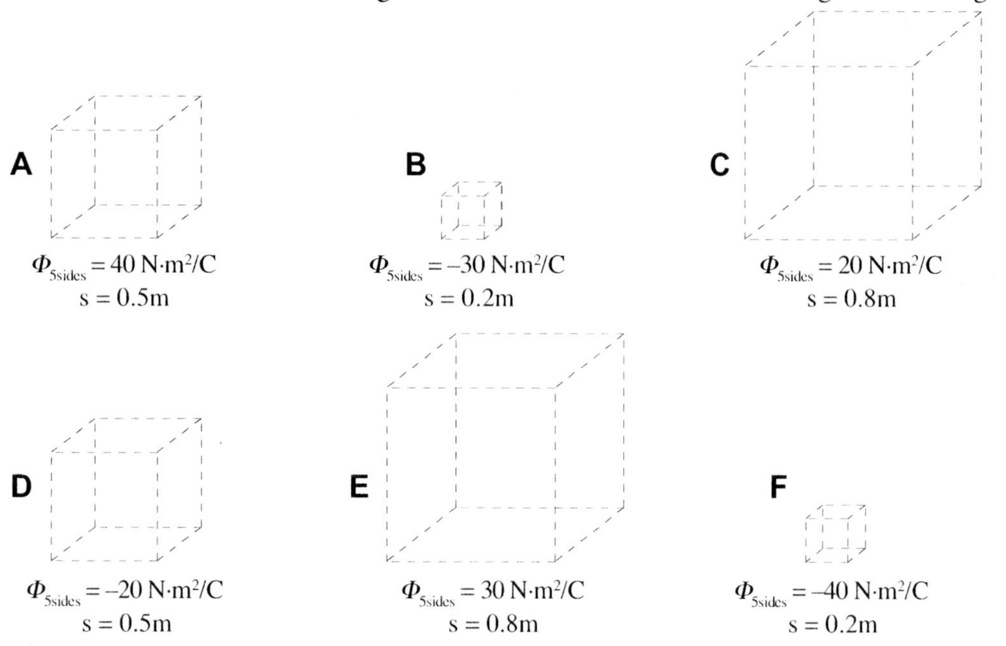

Rank the electric flux through the sixth face of the cube.

Greatest positive 1 ______ 2 ______ 3 ______ 4 ______ 5 ______ 6 ______ Greatest negative

OR, the electric flux through the sixth face will be the same (but not zero) for all six situations. ______

OR, the electric flux through the sixth face will be zero for all six situations. ______

OR, we cannot determine this ranking with the given information. ______

Carefully explain your reasoning.

How sure were you of your ranking? (circle one)

Basically Guessed					Sure				Very Sure
1	2	3	4	5	6	7	8	9	10

ET10-RT1: CHARGED ROD NEAR A SUSPENDED BAR MAGNET—TORQUE

A bar magnet is suspended by a string.

With the magnet held in place, a charged rod is brought close to the point of suspension of the magnet as shown.

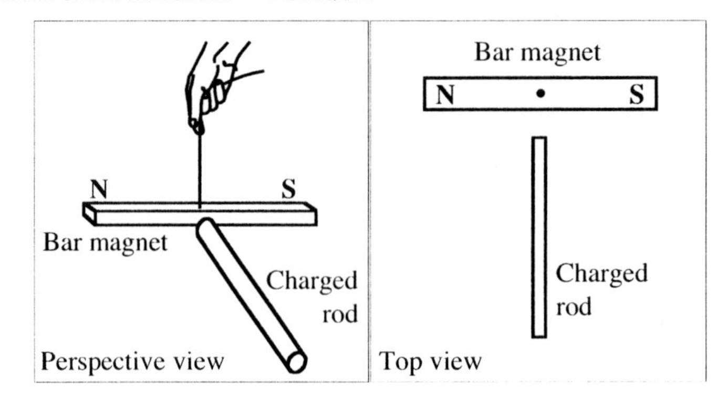

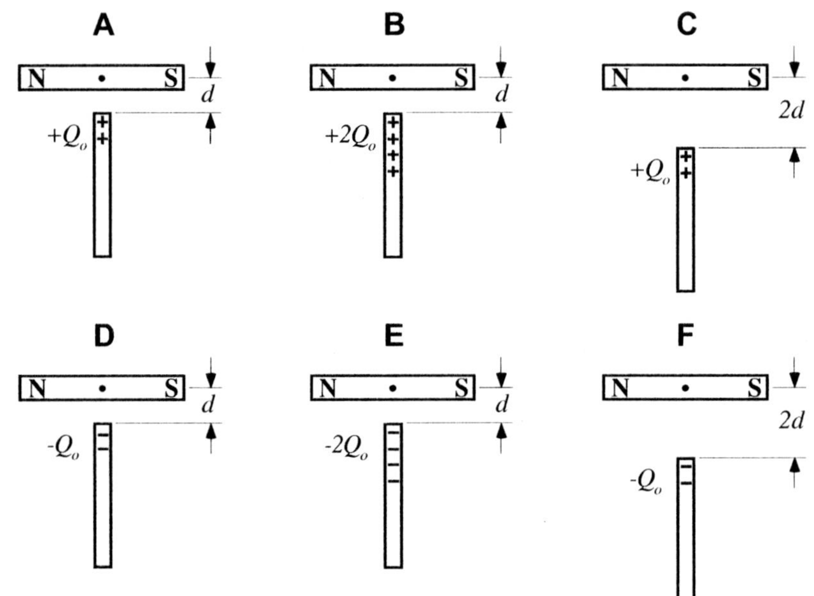

Rank the torque acting on the magnet about its center by the charged rod.

Greatest clockwise 1 ____ 2 ____ 3 ____ 4 ____ 5 ____ 6 ____ Greatest counterclockwise

OR, the torque is the same (but not zero) for all six situations. ____

OR, the torque is zero for all six situations. ____

OR, the ranking for the torques cannot be determined. ____

Carefully explain your reasoning.

How sure were you of your ranking? (circle one)

Basically Guessed				Sure					Very Sure
1	2	3	4	5	6	7	8	9	10

COMPARISON TASKS (CT)

ET1-CT1: CHARGES IN ELECTRIC FIELD—CHARGE

Shown below are the lines of equipotential for a region in which there is an electric field. Also shown are the paths followed by two positively-charged particles of equal mass. Both of the particles start with the same initial speed, but end with the different final speeds given.

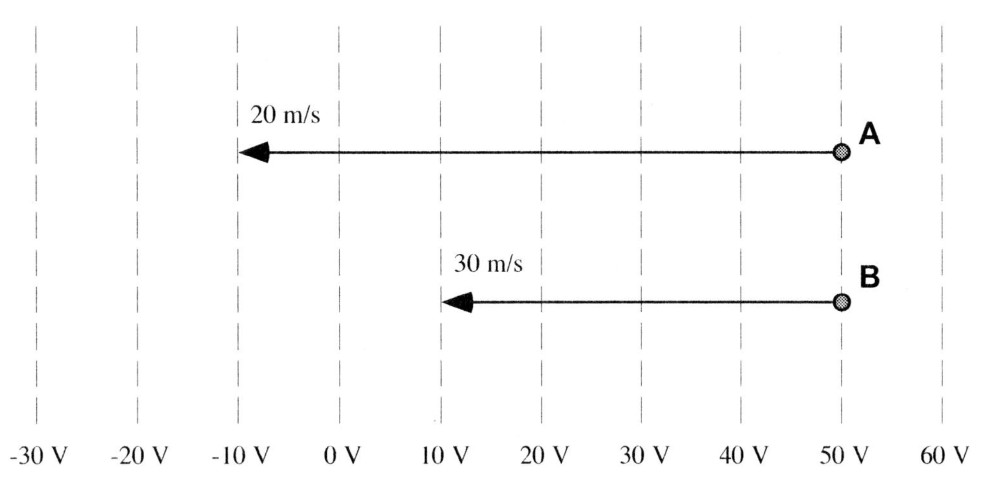

Is the charge of particle *A larger than, smaller than,* or *the same as* the charge of *B*? Explain.

ET3-CT1: STRAIGHT CHARGED ROD AND TWO POINT CHARGES—FORCE

A point charge $+Q$ is placed midway between a + 12 nC point charge and an insulating rod with a uniform charge distribution.

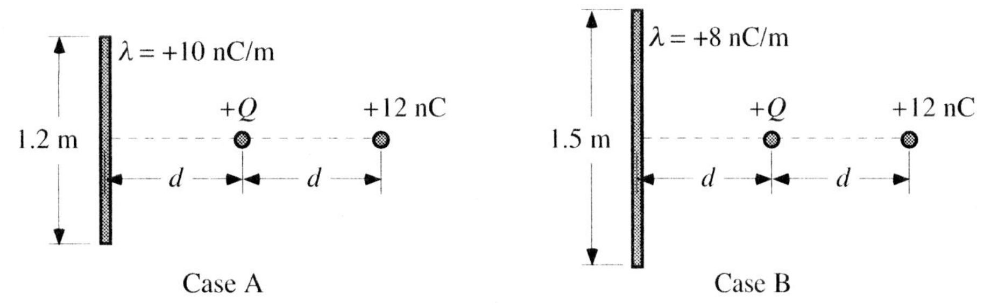

Case A Case B

Will the net force on $+Q$ in case A be *greater than, less than,* or *equal to* the force on $+Q$ in case B? Explain.

ET4-CT1: CART APPROACHING SPHERE—DISTANCE

In both cases below, identical carts carrying an electrically charged sphere is moving toward a second charged sphere that is fixed in place.

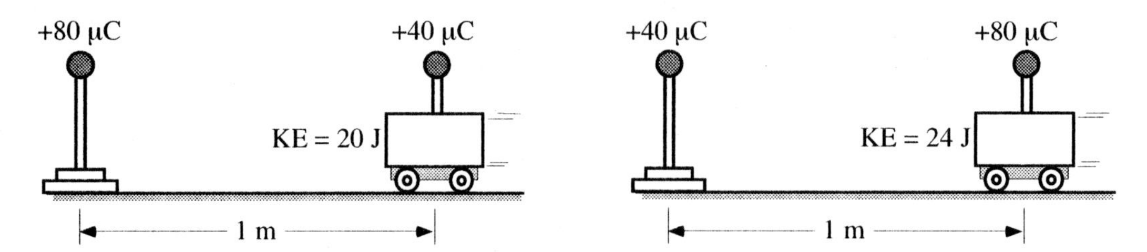

Will the cart in the situation on the left stop *closer to, farther from,* or *equally distant from* the fixed sphere compared to the cart in the situation at the right if we neglect friction? Explain.

ET5-CT1: POTENTIAL NEAR CHARGES—ELECTRIC FIELD

In the two cases below, a point midway between equal magnitude electric charges is identified. The signs of these charges are not given. The electric potential at the midpoint, in terms of V_0 the potential due to one charge alone, is also given.

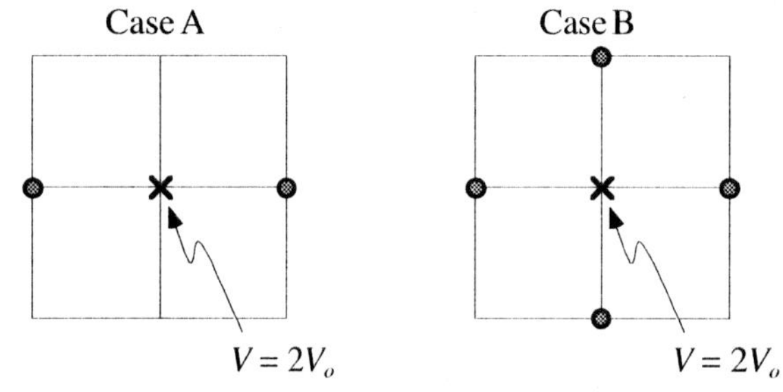

Is the magnitude of the electric field in Case A *greater than, less than,* or *equal to* the magnitude of the electric field in Case B? Explain.

ET5-CT2: POTENTIAL VS POSITION GRAPH II—ELECTRIC FIELD

Shown below is a graph of the electric potential versus position for a region of space. The labeled points indicate six locations within this region.

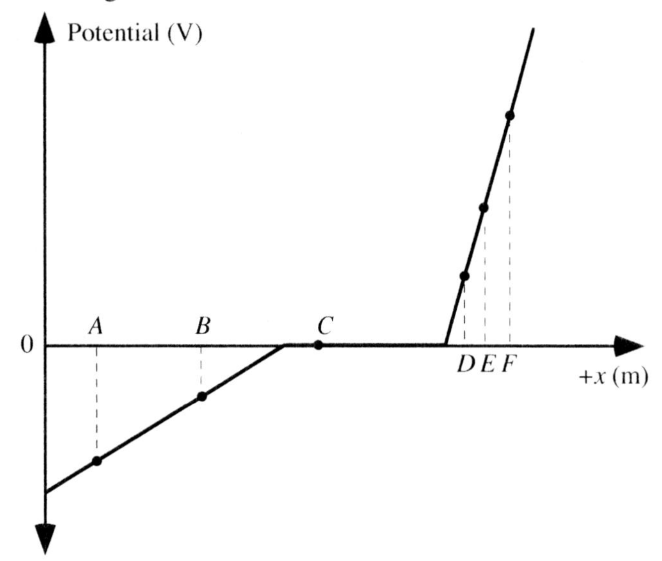

Compare the magnitude of the electric field at the following pairs of points.

Is the magnitude of the electric field:

1) **at *A greater than, less than,* or *equal to* the magnitude of the field at *B*? Explain.**

2) **at *A greater than, less than,* or *equal to* the magnitude of the field at *E*? Explain.**

3) **at *A greater than, less than,* or *equal to* the magnitude of the field at *C*? Explain.**

4) **at *D greater than, less than,* or *equal to* the magnitude of the field at *F*? Explain.**

ET6-CT1: THREE CHARGE SYSTEM—ELECTRIC POTENTIAL ENERGY AND WORK DONE

The figures below show the initial and final configurations of three charged particles. All charges have the same mass and are at rest in the initial and final configurations. They are initially arranged in identical equilateral triangles.

Choose the appropriate symbol (>, <, or =) to indicate:

(a) whether the electric potential energy of the initial configuration is greater than, less than, or equal to the final configuration and

(b) whether the work done on the system by an external agent in moving from initial to the final configuration was positive, negative, or zero.

	Initial Configuration		Final Configuration	Comparison (U_{init} (> or < or =) U_{final})	Work done on the system (W (> or < or =) 0)
Case 1	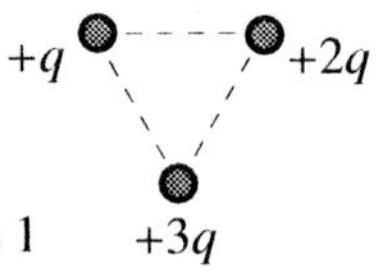 $+q$, $+2q$, $+3q$	⇒	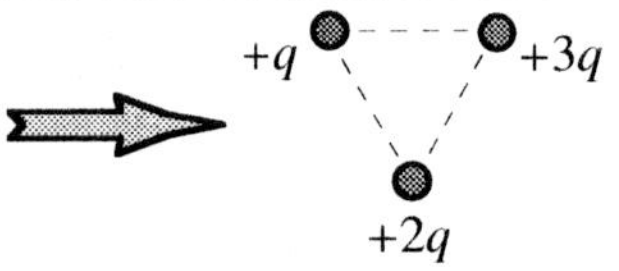 $+q$, $+3q$, $+2q$	U_{init} ____ U_{final}	W ____ 0
Case 2	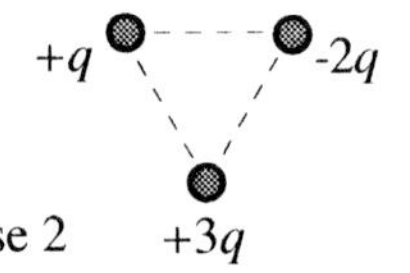 $+q$, $-2q$, $+3q$	⇒	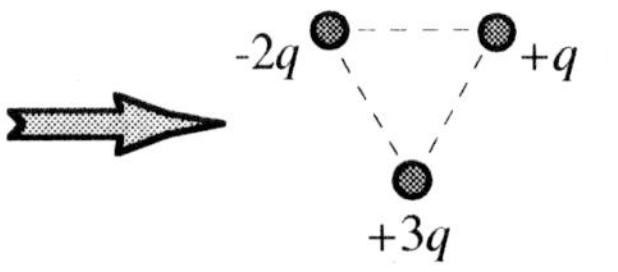 $-2q$, $+q$, $+3q$	U_{init} ____ U_{final}	W ____ 0
Case 3	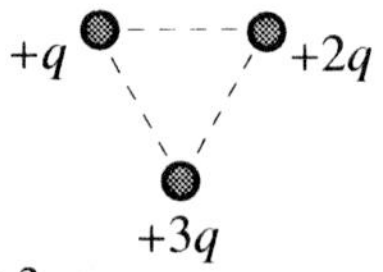 $+q$, $+2q$, $+3q$	⇒	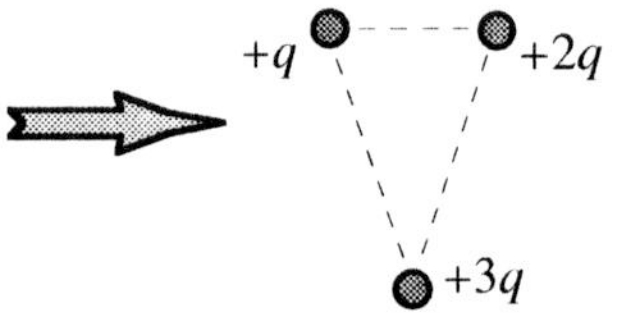 $+q$, $+2q$, $+3q$	U_{init} ____ U_{final}	W ____ 0
Case 4	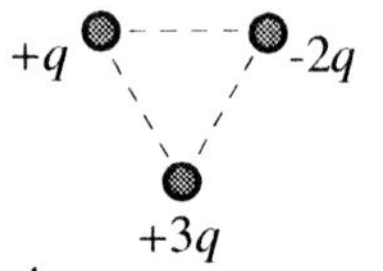 $+q$, $-2q$, $+3q$	⇒	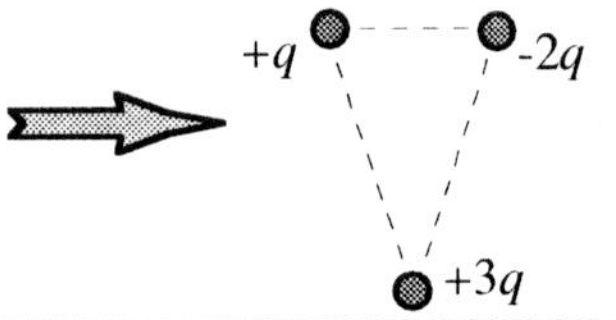 $+q$, $-2q$, $+3q$	U_{init} ____ U_{final}	W ____ 0
Case 5	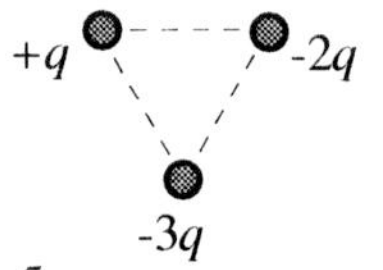 $+q$, $-2q$, $-3q$	⇒	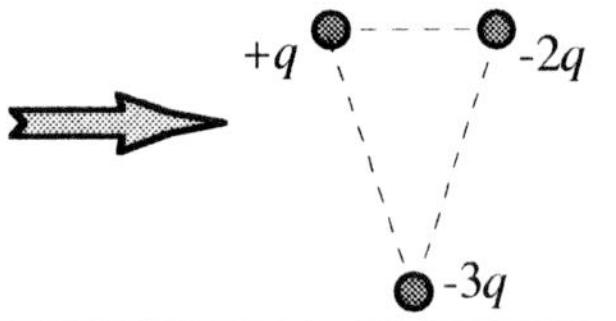 $+q$, $-2q$, $-3q$	U_{init} ____ U_{final}	W ____ 0
Case 6	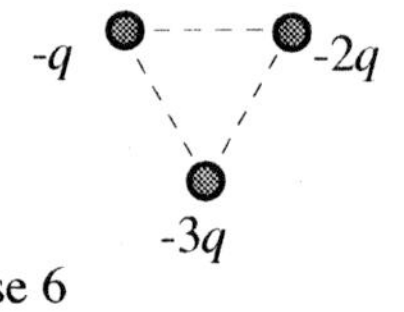 $-q$, $-2q$, $-3q$	⇒	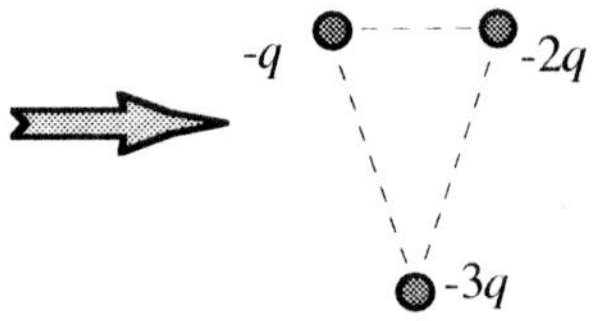 $-q$, $-2q$, $-3q$	U_{init} ____ U_{final}	W ____ 0

ET8-CT1: Points near Pair of Charges—Potential Difference

Two equal and opposite charges are fixed in space at the locations shown. Seven points are labeled *A* - *G* in the vicinity of these charges.

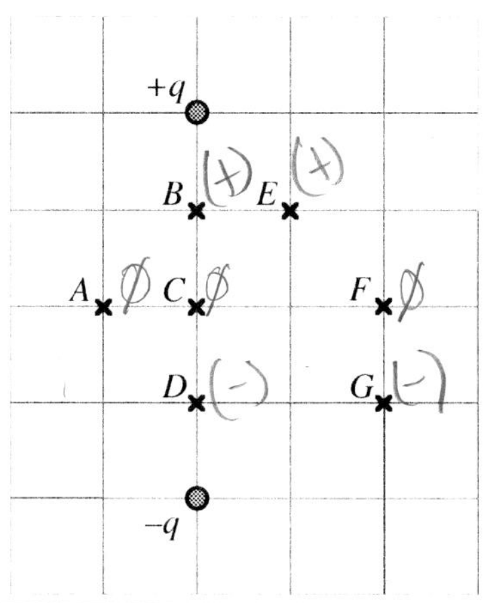

For each pair of points, decide whether the electric potential at point 1 is *greater than*, *less than*, or *equal to* the electric potential at point 2. Assume that the electric potential is zero far away from the charges.

Point 1	Point 2	Electric Potential (> or < or =)	Explanation
A	*C*	A = C	same change in distance (Δd)
A	*B*	A < B	B is closer to the (+) charge
A	*D*	A > D	
A	*F*	A = F	
A	*E*	A < E	
B	*C*	B > C	
B	*D*	B > D	
B	*E*	B > E	
F	*G*	F > G	
D	*E*	D < E	

QUALITATIVE REASONING TASKS (QRT)

ET1-QRT1: BREAKING A CHARGED INSULATING BLOCK—CHARGE AND CHARGE DENSITY

The block of insulating material shown below has a length L_o and a volume V_o. An overall charge Q_o is spread uniformly throughout the volume of the block to give a volume charge density ρ_o and a linear charge density λ_o (in the direction of the measured length L_o).

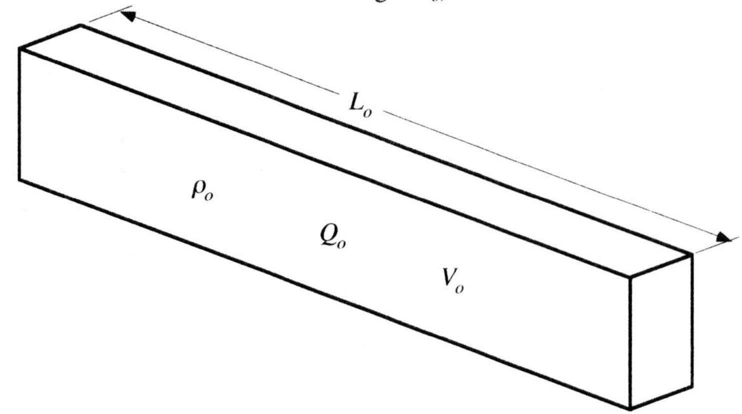

Three possible ways to split the block into unequal pieces are represented below. In each case, the larger piece has a volume $2V_o/3$ and the smaller piece has a volume $V_o/3$.

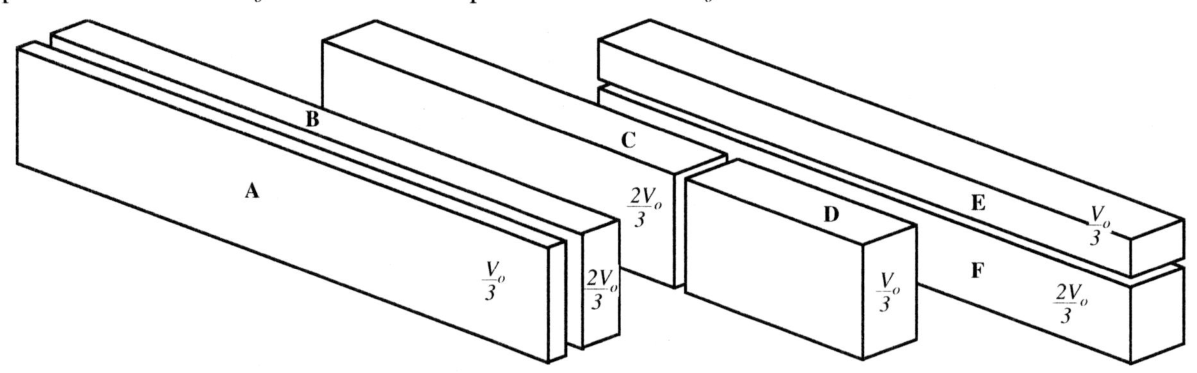

Fill in the table with the quantities indicated. Express your answers in terms of the variables Q_o, λ_o, and ρ_o.

	Charge	Charge per unit length	Charge per unit volume
Original block	Q_o	λ_o	ρ_o
Piece A			
Piece B			
Piece C			
Piece D			
Piece E			
Piece F			

ET1-QRT2: CHARGED INSULATING BLOCKS—ORIGINAL BLOCK

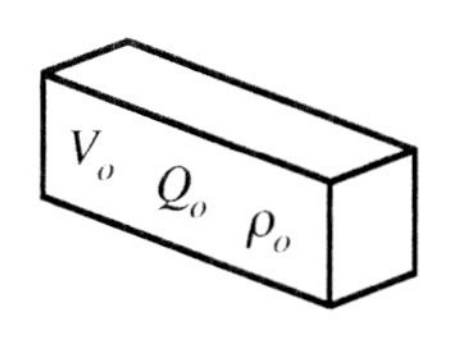

The block of insulating material shown at right has a volume V_o. An overall charge Q_o is spread uniformly throughout the volume of the block so that the block has a charge density ρ_o.

Four charged insulating blocks are shown below. For each block, the volume is given as well as *either* the charge or the charge density of the block.

Suppose that the blocks below were cut in half. From which block or blocks might the block above have been taken? Explain.

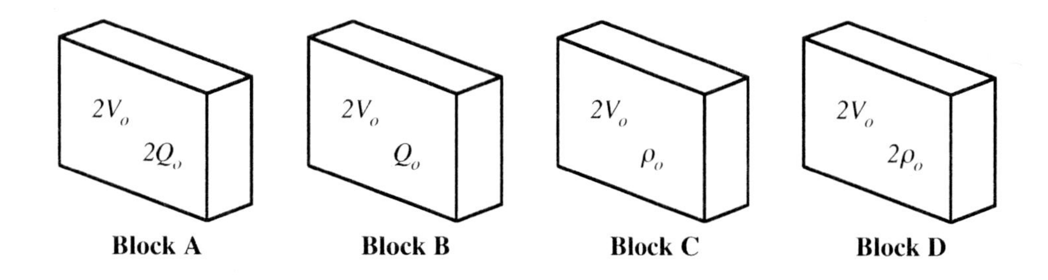

ET1-QRT3: CHARGED INSULATING BLOCKS—CHARGE AND CHARGE DENSITY

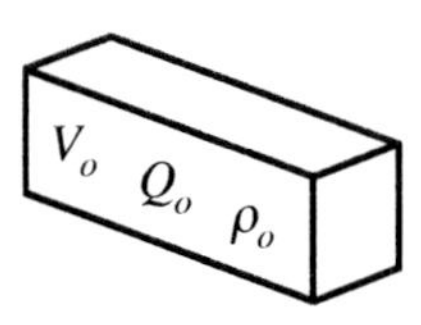

The block of insulating material shown at the right has a volume V_o. An overall charge Q_o is spread uniformly throughout the volume of the block so that the block has a charge density ρ_o.

Four charged insulating blocks are shown below. For each block, the volume is given as well as *either* the charge or the charge density of the block.

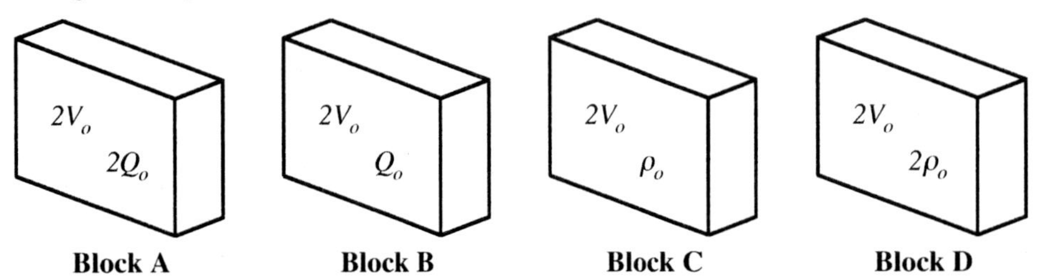

Fill in the table below with the quantities indicated. Express your answers in terms of the variables V_o, Q_o, and ρ_o.

	Charge	Charge per unit volume
Original block	Q_o	ρ_o
Block A		
Block B		
Block C		
Block D		

ET1-QRT4: CHARGED INSULATING ROD—CHARGE AND CHARGE DENSITY

The insulating rod shown at top right has a length L and a radius r. The rod has an overall charge Q_o spread uniformly throughout its volume to give a charge density ρ_o and a linear charge density λ_o.

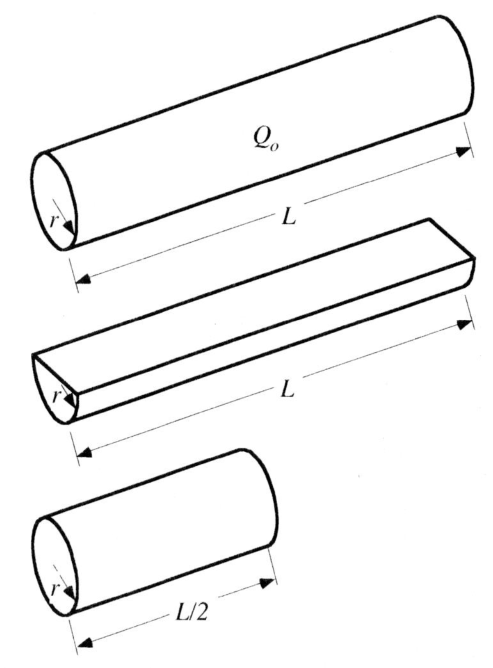

Below the rod are drawings of (1) half the rod after it is sliced lengthwise, and (2) half the rod after it is sliced across its middle or crosswise.

Fill in the table below giving the quantities indicated for the two half rods. Express your answers in terms of the values in the first row.

	Charge	**Charge per unit length**	**Charge per unit volume**
Original rod	Q_o	λ_o	ρ_o
1. Rod sliced lengthwise			
2. Rod sliced crosswise			

ET1-QRT5: THREE CONDUCTING SPHERES—CHARGE

Two conducting spheres rest on insulating stands. Sphere *B* is smaller than sphere *A*. Both spheres are initially uncharged and they are touching.

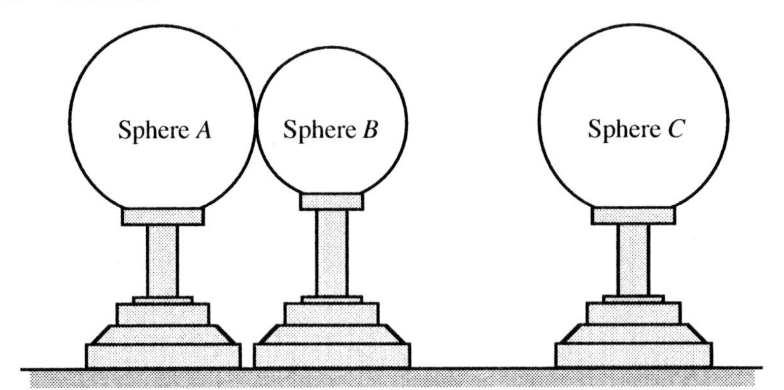

A third conducting sphere, *C*, has a positive charge. It is brought close to (but not touching) sphere *B* as shown.

Is the net charge on Sphere *A* at this time *positive, negative*, or *zero*?

Is the net charge on Sphere *B* at this time *positive, negative*, or *zero*?

Is the magnitude of the net charge on Sphere *A* *greater than, less than*, or *equal to* the magnitude of the net charge on Sphere *B*?

Is the magnitude of the surface charge density on Sphere *A* *greater than, less than*, or *equal to* the magnitude of the surface charge density on Sphere *B*?

Sphere *B* is now moved to the right so that it touches Sphere *C*. As a result of this move:

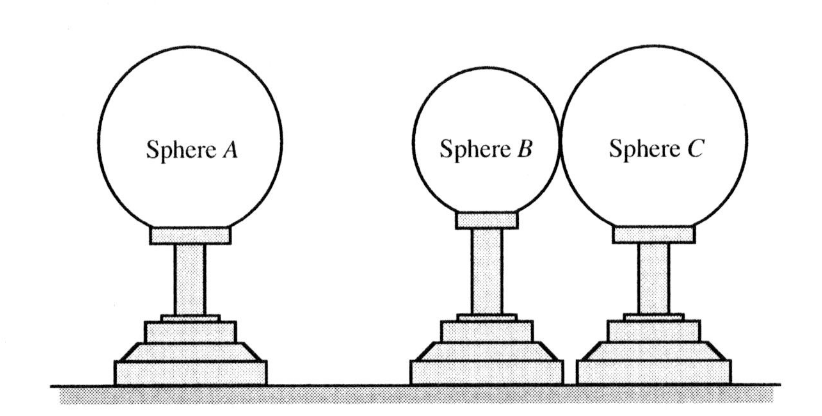

Does the magnitude of the net charge on Sphere *A* *increase, decrease*, or *remain the same*?

Does the magnitude of the net charge on Sphere *C* *increase, decrease*, or *remain the same*?

ET3-QRT1: Two Unequal Charges—Force

Shown below are two charged particles that are fixed in place. The magnitude of the charge Q is greater than the magnitude of the charge q. A third charge is now placed at one of the points A – E. The net force on this charge due to q and Q is zero.

1. Both q and Q are positive.
   At which point A – E is it possible that the third charge was placed? Explain.

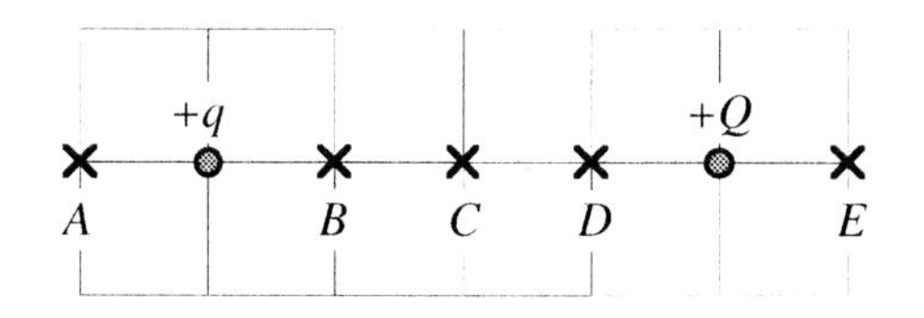

2. Charge q is positive and charge Q is negative.
   At which point A – E is it possible that the third charge was placed? Explain.

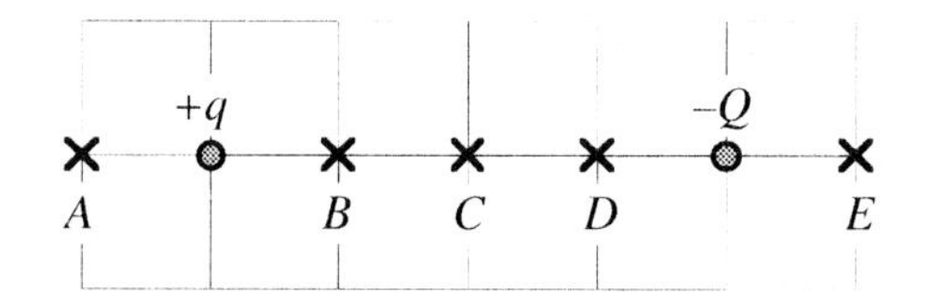

3. Charge q is negative and charge Q is positive.
   At which point A – E is it possible that the third charge was placed? Explain.

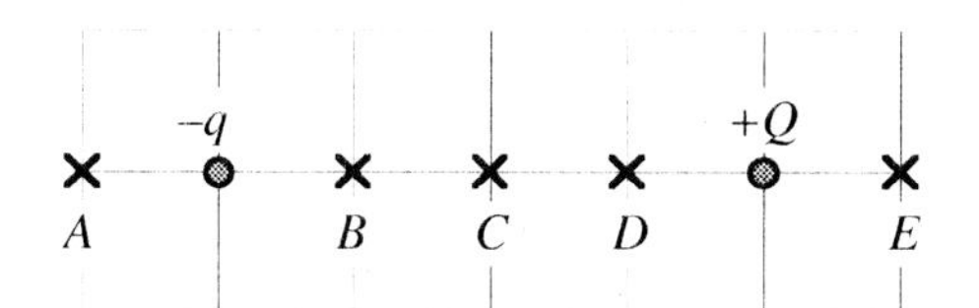

4. Both q and Q are negative.
   At which point A – E is it possible that the third charge was placed? Explain.

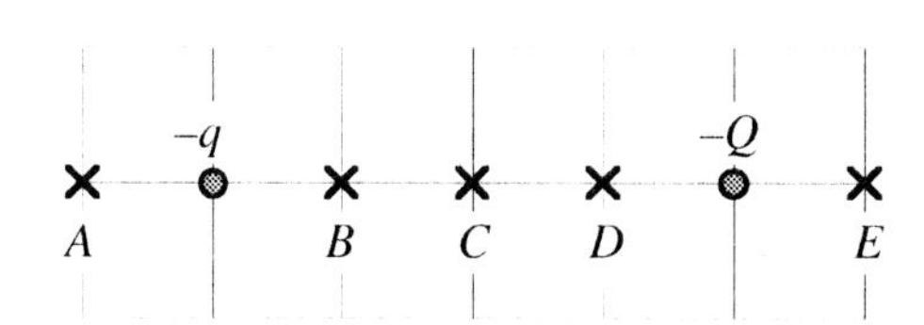

ET3-QRT2: THREE CHARGES IN A LINE—FORCE

Two charged particles, *A* and *B*, are fixed in place. A third charge, *C*, is fixed in place to the right of charge *B* at twice the distance between *A* and *B*. All charges are the same magnitude.

In the chart to the left below, use arrows to indicate the direction of the net force on charge *C* due to charges *A* and *B*. If the force is zero, state that explicitly.

In the chart on the right below, use arrows to indicate the direction of the net force on charge *B* due to charges *A* and *C*. If the force is zero, state that explicitly.

A	*B*	*C*	$\Sigma\vec{F}$ on charge *C*
+	+	+	Direction:
+	+	−	Direction:
+	−	+	Direction:
+	−	−	Direction:
−	+	+	Direction:
−	+	−	Direction:
−	−	+	Direction:
−	−	−	Direction:

A	*B*	*C*	$\Sigma\vec{F}$ on charge *B*
+	+	+	Direction:
+	+	−	Direction:
+	−	+	Direction:
+	−	−	Direction:
−	+	+	Direction:
−	+	−	Direction:
−	−	+	Direction:
−	−	−	Direction:

ET3-QRT3: THREE CHARGES IN A LINE—FORCE

Two charged particles, *A* and *B*, are fixed in place. A third charge, *C*, is fixed in place to the right of charge *B* at twice the distance between *A* and *B*. The magnitudes of the charges of *A, B,* and *C* are all different.

A *B* *C*

For each of the following combinations of charged particles, determine if it is possible for the net electric force on each charge due to the other charges to be zero.

A	*B*	*C*	$\Sigma\vec{F}$ on charge *A*	$\Sigma\vec{F}$ on charge *B*	$\Sigma\vec{F}$ on charge *C*
+	+	+	Must be nonzero ☐ Possibly zero ☐	Must be nonzero ☐ Possibly zero ☐	Must be nonzero ☐ Possibly zero ☐
+	+	−	Must be nonzero ☐ Possibly zero ☐	Must be nonzero ☐ Possibly zero ☐	Must be nonzero ☐ Possibly zero ☐
+	−	+	Must be nonzero ☐ Possibly zero ☐	Must be nonzero ☐ Possibly zero ☐	Must be nonzero ☐ Possibly zero ☐
+	−	−	Must be nonzero ☐ Possibly zero ☐	Must be nonzero ☐ Possibly zero ☐	Must be nonzero ☐ Possibly zero ☐
−	+	+	Must be nonzero ☐ Possibly zero ☐	Must be nonzero ☐ Possibly zero ☐	Must be nonzero ☐ Possibly zero ☐
−	+	−	Must be nonzero ☐ Possibly zero ☐	Must be nonzero ☐ Possibly zero ☐	Must be nonzero ☐ Possibly zero ☐
−	−	+	Must be nonzero ☐ Possibly zero ☐	Must be nonzero ☐ Possibly zero ☐	Must be nonzero ☐ Possibly zero ☐
−	−	−	Must be nonzero ☐ Possibly zero ☐	Must be nonzero ☐ Possibly zero ☐	Must be nonzero ☐ Possibly zero ☐

ET3-QRT4: THREE CHARGES IN A LINE—FORCE

Two charged particles, *A* and *B*, are fixed in place. A third charge, *C*, is fixed in place to the right of charge *B* at twice the distance between *A* and *B*. There is no net force on charge *C* due to charges *A* and *B*.

A B C
+ − +
$\Sigma\vec{F} = 0$

Indicate whether each of the following statements is *true, false,* or *cannot be determined.*

	Statement	True	False	Cannot be Determined
1.	**Charge *A* has a greater magnitude than charge *C*.**			
2.	**Charge *A* has a greater magnitude than charge *B*.**			
3.	**Charge *C* has a greater magnitude than charge *B*.**			
4.	**Charge *A* has the same magnitude as charge *C*.**			
5.	**Charge *A* has the same magnitude as charge *B*.**			
6.	**Charge *C* has the same magnitude as charge *B*.**			

Two charged particles, *A* and *B*, are fixed in place. A third charge, *C*, is placed to the right of charge *B* at twice the distance between *A* and *B*. There is no net force on charge *B* due to charges *A* and *C*.

A B C
+ − +
$\Sigma\vec{F} = 0$

Indicate whether each of the following statements is *true, false,* or *cannot be determined.*

	Statement	True	False	Cannot be Determined
7.	**Charge *A* has a greater magnitude than charge *C*.**			
8.	**Charge *A* has a greater magnitude than charge *B*.**			
9.	**Charge *C* has a greater magnitude than charge *B*.**			
10.	**Charge *A* has the same magnitude as charge *C*.**			
11.	**Charge *A* has the same magnitude as charge *B*.**			
12.	**Charge *C* has the same magnitude as charge *B*.**			

ET3-QRT5: STRAIGHT CHARGED ROD AND TWO POINT CHARGES—FORCE

A point charge labeled $+Q$ is sitting midway between a +10 nC point charge and a rod of length 1 m with a uniform charge distribution.

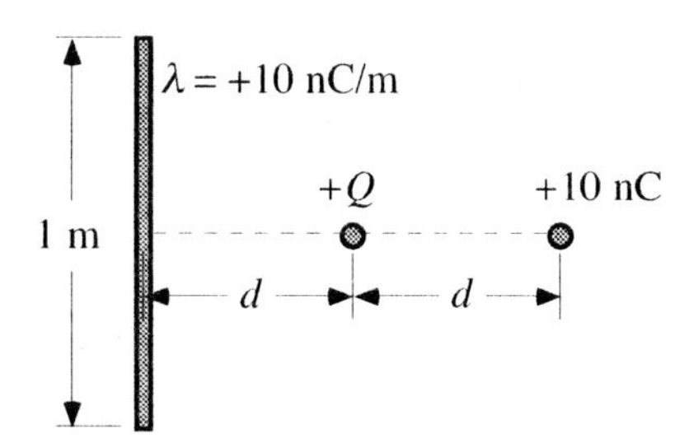

Explain how each of the modifications to this initial situation described below will affect the net force acting on $+Q$ in the original situation.

(a) The rod is moved to the left, increasing its distance from $+Q$.

(b) The length of the rod is reduced while keeping the same total charge.

(c) The +10 nC point charge is replaced by a +12 nC point charge.

(d) The charge on the rod is changed to negative.

(e) The charge density on the rod is reduced +8 nC/m.

(f) The length of the rod is doubled while keeping the same charge density.

(g) The rod is replaced by a +10nC point charge placed on the dashed line a distance d to the left of $+Q$.

ET3-QRT6: CHARGE NEAR EQUIPOTENTIAL SURFACES—FORCE DIRECTION

The dotted lines in the diagram represent cross-sections of flat electric equipotential surfaces or sheets. The distance between adjacent equipotentials is 10 centimeters.

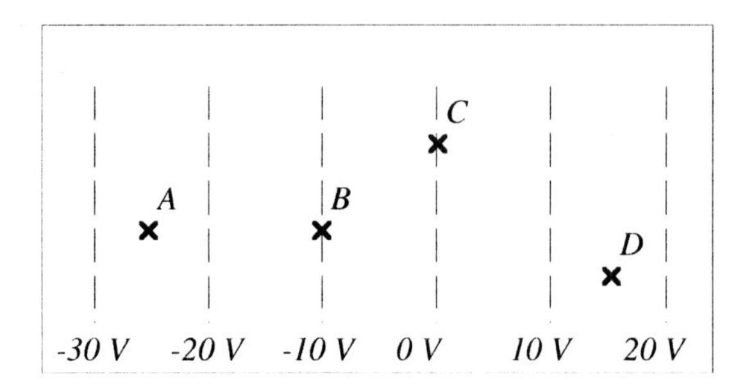

Question	Answer
1) What is the direction of the force on an electron if it is placed at rest at ***A*?** (If the force is zero state that explicitly.)	
2) What is the direction of the force on an electron if it is placed at rest at ***B*?** (If the force is zero state that explicitly.)	
3) What is the direction of the force on an electron if it is placed at rest at ***C*?** (If the force is zero state that explicitly.)	
4) What is the direction of the force on an electron if it is placed at rest at ***D*?** (If the force is zero state that explicitly.)	
5) At which location(s) is the force on an electron the greatest?	
6) What is the direction of the force on a proton if it is placed at rest at ***B*?** (If the force is zero state that explicitly.)	
7) What is the direction of the force on a proton if it is placed at rest at ***D*?** (If the force is zero state that explicitly.)	
8) At which location(s) is the force on a proton the greatest?	

ET3-QRT7: Force Direction on Charges in an Equilateral Triangle—Force

Three charges are fixed at the vertices of each equilateral triangle shown below. All charges have the same magnitude.

Determine the direction of the net electric force acting on each charge due to the other two charges in the same triangle. Answer by using letters A through L representing directions from the choices below.

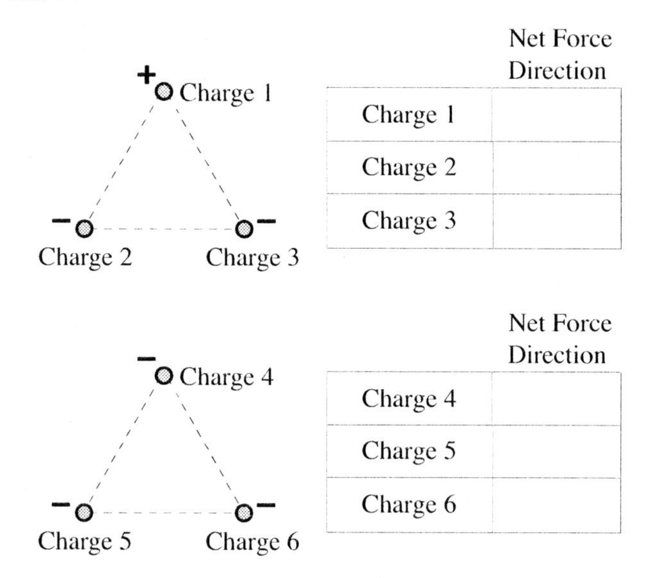

	Net Force Direction
Charge 1	
Charge 2	
Charge 3	

	Net Force Direction
Charge 4	
Charge 5	
Charge 6	

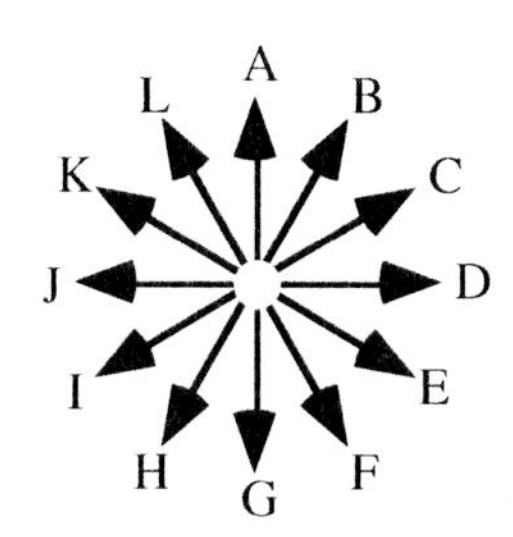

ET3-QRT8: Force Direction on Charges in a Right Triangle—Force

Three charges are fixed at the vertices of each right isosceles triangle shown below. All charges have the same magnitude.

Determine the direction of the net electric force acting on each charge due to the other two charges in the same triangle. Answer by using letters A through H representing directions from the choices below. If the angle is between two directions, indicate both directions such as AB for a direction between A and B.

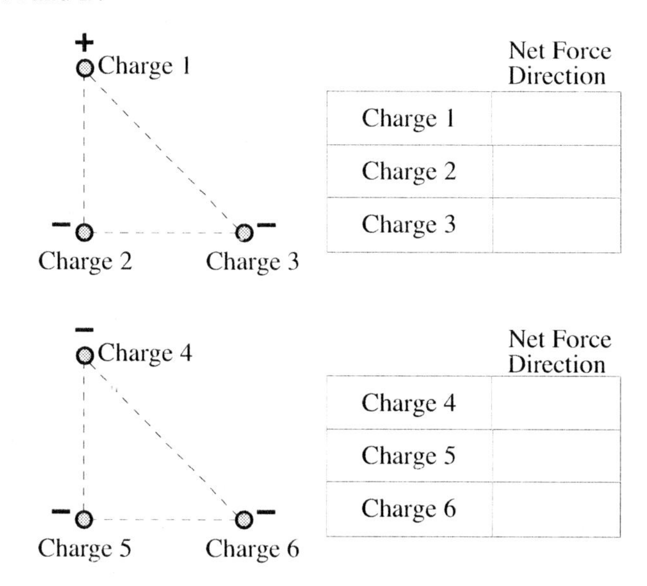

	Net Force Direction
Charge 1	
Charge 2	
Charge 3	

	Net Force Direction
Charge 4	
Charge 5	
Charge 6	

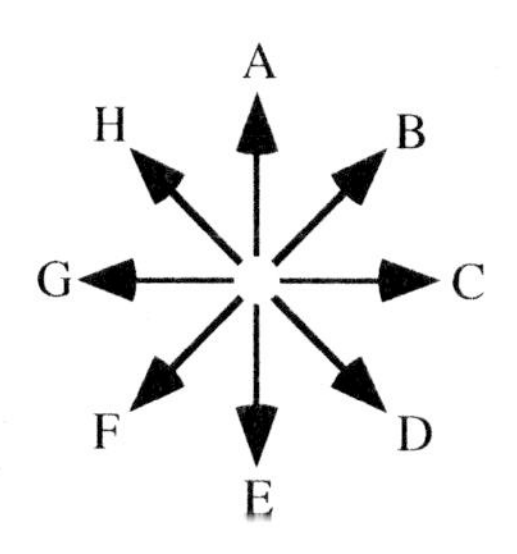

ET3-QRT9: FORCE DIRECTION ON CHARGES IN A SQUARE—FORCE

Four charges are fixed at the vertices of each square shown below. All charges have the same magnitude.

Determine the direction of the net electric force acting on each charge due to the other three charges in the same square. Answer by using letters A through H representing directions from the choices given below. If the angle is between two labeled directions, indicate those two directions (AB for a direction between A and B, for example).

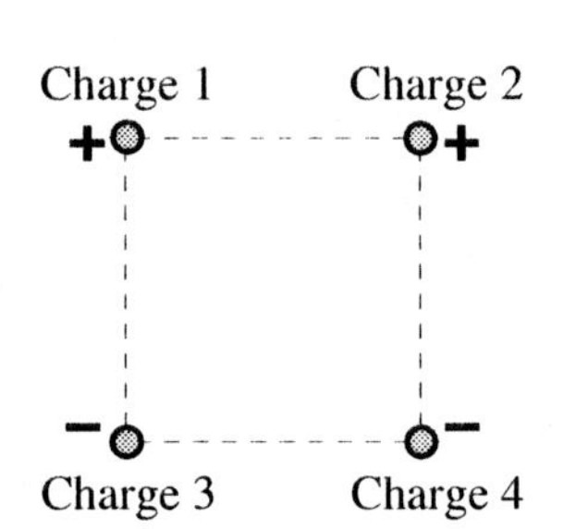

	Net Force Direction
Charge 1	
Charge 2	
Charge 3	
Charge 4	

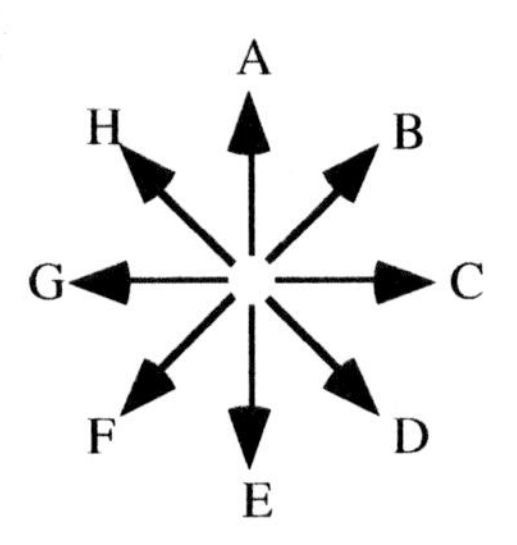

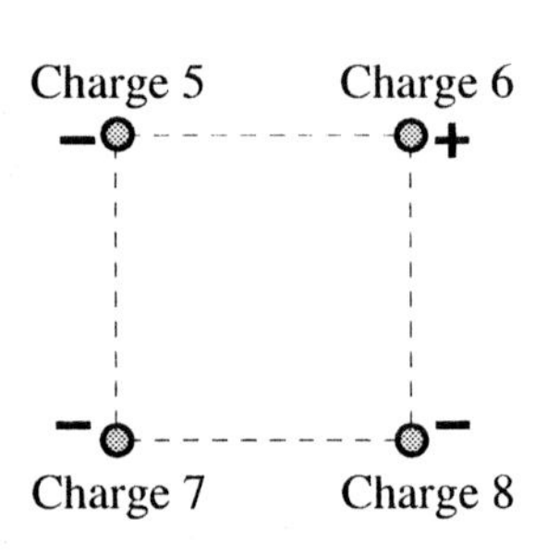

	Net Force Direction
Charge 5	
Charge 6	
Charge 7	
Charge 8	

ET3-QRT10: Two Charges—Force on Each

In each case shown below, two charges are exerting forces on each other.

For each case, draw a vector of appropriate length and direction representing the electric force acting on each charge due to the other charge. Draw the vector representing the force on the left charge above that charge; draw the vector representing the force on the right charge below that charge (see example). For each diagram use the same scale as the example.

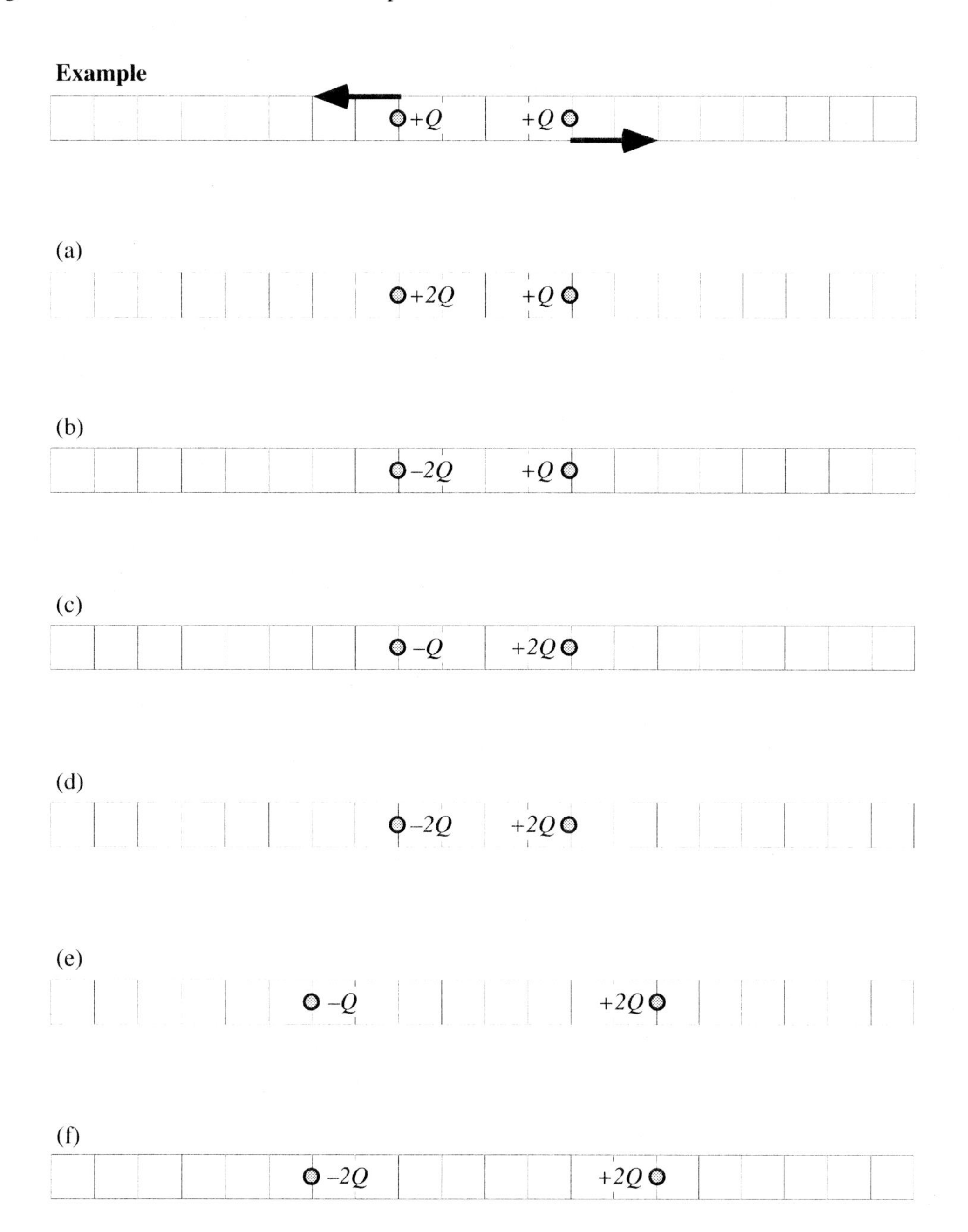

ET6-QRT1: TWO CHARGED OBJECTS—WORK AND ENERGY

Shown below are the positions of two small objects with identical masses and different charges. The smaller charge is fixed in place while the $+4q$ charge with mass m is free to move. At time t_1, the charges are at rest and separated by a distance d. At a later time t_2, the charges are separated by a distance $2d$.

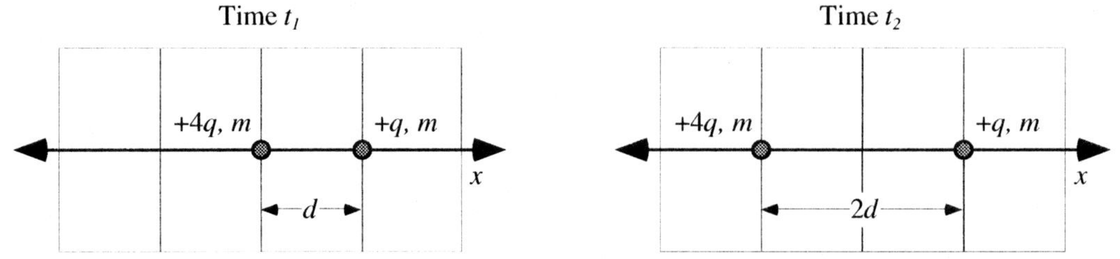

A. **Is the work done by the $+q$ charge on the $+4q$ charge during the time interval from t_1 to t_2 *positive, negative,* or *zero*?** If the work done cannot be determined with the given information, state that explicitly. **Explain.**

B. **Will the electric potential energy of the system of two charges *increase, decrease,* or *remain the same* during the time interval from t_1 to t_2?** If the change in electric potential energy cannot be determined with the given information, state that explicitly. **Explain.**

C. **Will the kinetic energy of the system of the two charges *increase, decrease,* or *remain the same* during the time interval from t_1 to t_2?** If the change in kinetic energy cannot be determined with the given information, state that explicitly. **Explain.**

D. **If the $+4q$ charged object with mass m is replaced with a $+4q$ charged object with a mass of $2m$, will the quantities in parts A, B, and C *be larger, smaller* or *the same as* determined in parts A, B, and C? Explain.**

E. **If the $+4q$ charged object with mass m is replaced with a $+12q$ charged object with the same mass m, will the quantities in parts A, B, and C *be larger, smaller,* or *the same as* determined in parts A, B, and C? Explain.**

ET5-QRT1: POTENTIAL VS POSITION GRAPHS—ELECTRIC FIELD

The graphs below of electric potential versus position are for regions in which there may be electric fields.

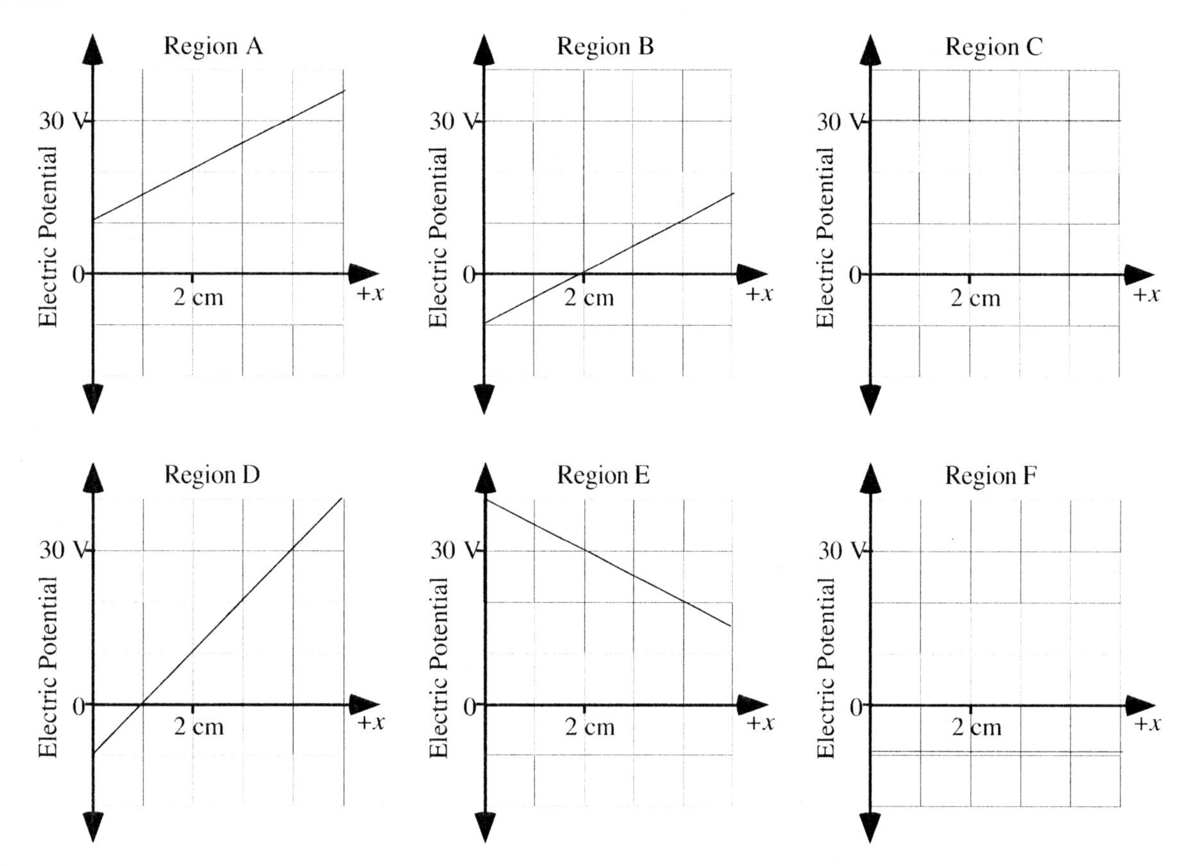

On the graphs above, draw a second line that is consistent with each of the following modifications.

(a) **The direction of the electric field in Region *A* is reversed while its magnitude is unchanged and the potential at $x = 0$ cm remains the same.**

(b) **The electric field in Region *B* remains the same but the potential at $x = 2$ cm increases to 10 volts.**

(c) **The electric field in Region *C* remains the same but the potential is cut in half at $x = 2$ cm.**

(d) **The magnitude of the electric field in Region *D* is increased keeping the same direction and the potential at $x = 2$ cm remains the same.**

(e) **The direction of the electric field in Region *E* is reversed and its magnitude increases but the potential at $x = 4$ cm remains the same.**

(f) **The electric field in Region *F* remains the same but the potential at $x = 2$ cm is 20 volts.**

ET5-QRT2: CHARGED INSULATING RODS—ELECTRIC FIELD

The figure shows portions of two very long insulating rods. The top rod has a uniform negative linear charge density of magnitude 12 μC/m, and the bottom rod has a uniform negative linear charge density of magnitude 4 μC/m.

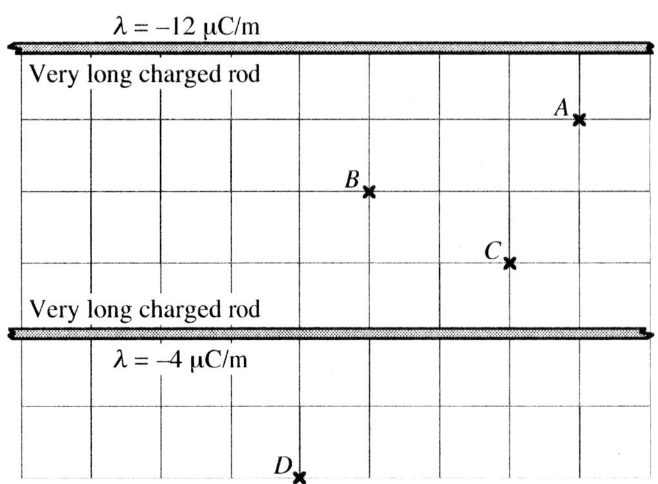

In the diagram, indicate the direction of the electric field at points *A, B, C,* and *D*. If the electric field is zero at any of these points, state that explicitly.

ET10-QRT1: GRAPH OF CHARGE VS ELECTRIC POTENTIAL—CAPACITANCE

Given below is a graph of the charge versus the electric potential for a conductor shaped like a U.

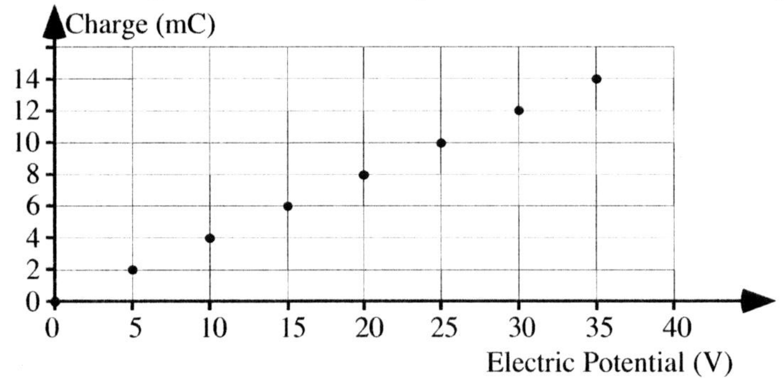

What is the capacitance of the U-shaped conductor at 30 V? Explain.

What is the capacitance of the U-shaped conductor at 10 V? Explain.

What is the capacitance of the U-shaped conductor at 0 V? Explain.

ET9-QRT1: CHARGE WITHIN A HOLLOW CONDUCTOR—ELECTRIC FLUX

A very small, charged, metal sphere is placed inside a thin conducting spherical shell of radius B without touching it. Two Gaussian spheres of radius A and C are used to find the net electric flux inside and outside of the shell.

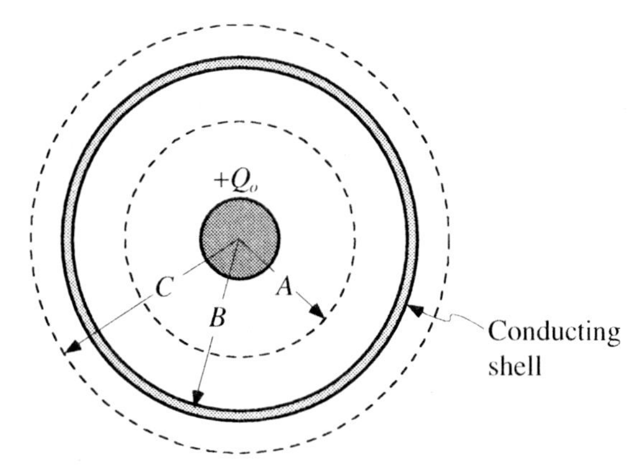

a) Suppose the two Gaussian surfaces have the same electric flux. **What is the charge on the inner and outer surfaces of the conducting shell? Explain.**

b) Now suppose the electric flux through the outer Gaussian surface is three times that through the inner Gaussian surface. **What is the charge on the inner and outer surfaces of the conducting shell? Explain.**

c) If the electric flux through the outer Gaussian surface is zero, what is the electric flux through the inner Gaussian surface? Explain.

d) If the electric flux through the outer Gaussian surface is zero, what is the charge on the inner and outer surfaces of the conducting shell? Explain.

LINKED MULTIPLE CHOICE TASKS (LMCT)

ET3-LMCT1: CHARGES ARRANGED IN A TRIANGLE—FORCE

Three charges are arranged in an isosceles triangle as shown.

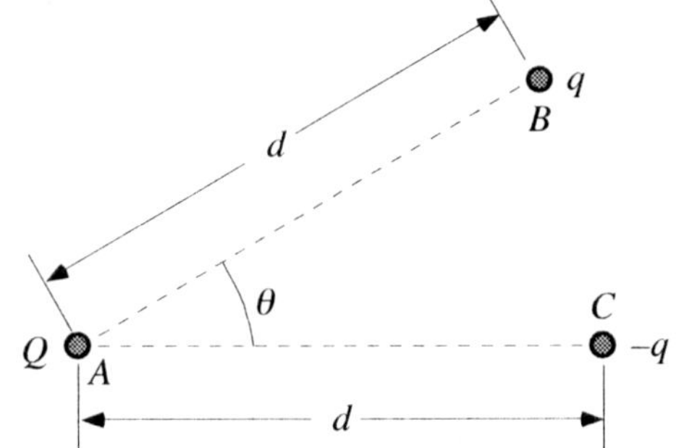

For each of the changes described, decide whether the magnitude of the net electric force on charge *A* will *increase, decrease, remain the same,* or *the effect cannot be determined.* Consider each change to be the only change from the situation in the diagram.

Change	Increase	Decrease	Remain the Same	Cannot be Determined
1. Change sign of charge *C* from –*q* to *q*				
2. Change sign of charge *B* from *q* to -*q*				
3. Change sign of charge *A* from *Q* to -*Q*				
4. Increase the angle *θ* by 5°				
5. Change the sign of both *A* and *C*				
6. Halve the distance between *A* and *C*				
7. Halve the distance between *A* and *B*				

ET3-LMCT2: SYSTEM OF CHARGES—ELECTRIC FORCE ON A CHARGE

The figure shows a system of three positive point charges.

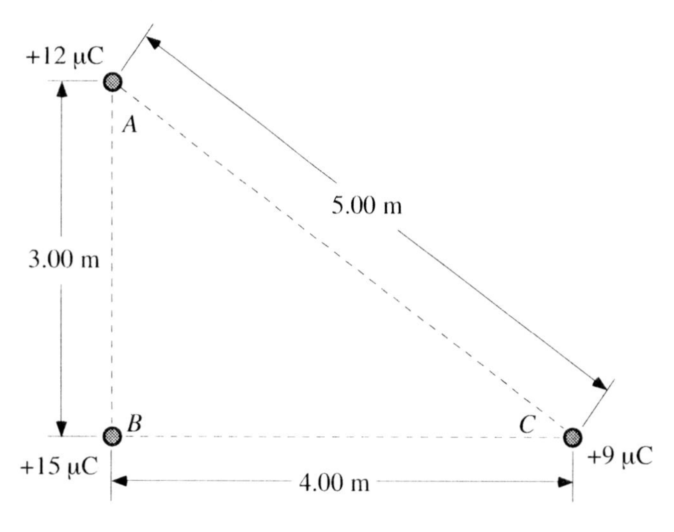

From the following choices identify how each change below (1-8) will affect the electric force on *B*.

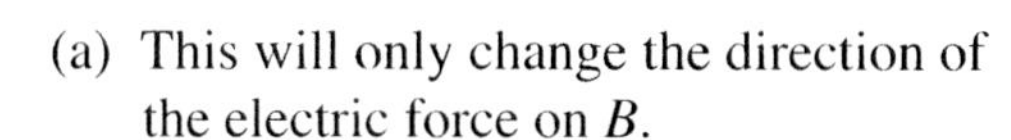

(a) This will only change the direction of the electric force on *B*.

(b) This will only increase the magnitude of the electric force on *B*.

(c) This will only decrease the magnitude of the electric force on *B*.

(d) This will increase the magnitude and change the direction of the electric force on *B*.

(e) This will decrease the magnitude and change the direction of the electric force on *B*.

(f) This will not affect the electric force on *B*.

All of these changes are modifications to the initial situation shown in the diagram.

1) **The charge on B is doubled.** ______

2) **The sign of the charge on B is changed to the opposite sign.** ______

3) **The charge on A and C are doubled.** ______

4) **The sign of the charges on *A* and *B* are both changed to the opposite sign.** ______

5) **The charge on A is doubled.** ______

6) **A is moved so the distance between A and B is halved.** ______

7) **C is moved so the distance between C and B is doubled.** ______

8) **The charge on every particle is halved.** ______

ET3-LMCT3: STRAIGHT CHARGED ROD AND TWO POINT CHARGES—FORCE

A point charge $+Q$ is placed midway between a +10 nC point charge and a rod with a uniform charge distribution.

Choose from this list how each of the modifications (1-5) below will affect the net force acting on $+Q$.

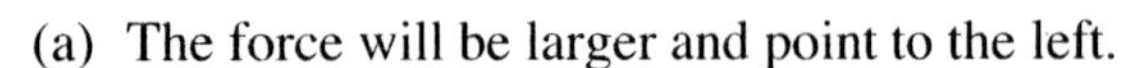

(a) The force will be larger and point to the left.

(b) The force will be larger and point to the right.

(c) The force will be smaller and point to the left.

(d) The force will be smaller and point to the right.

(e) The force will have the same magnitude but opposite direction.

(f) The force will be the same as in the initial situation above.

(g) The force will be zero.

All of these modifications are changes to the initial situation shown in the diagram.

1) The charge density on the rod is reduced. _______

2) The length of the rod is reduced by one third without reducing the total charge. _______

3) The magnitude of $+Q$ is decreased. _______

4) The +10 nC charge is replaced with a +15 nC charge. _______

5) The +10 nC charge is replaced with a -10 nC charge. _______

ET3-LMCT4: SPHERE AND A POINT CHARGE—FORCE

A positive point charge is placed a distance d away from the closest surface of a neutral metal sphere.

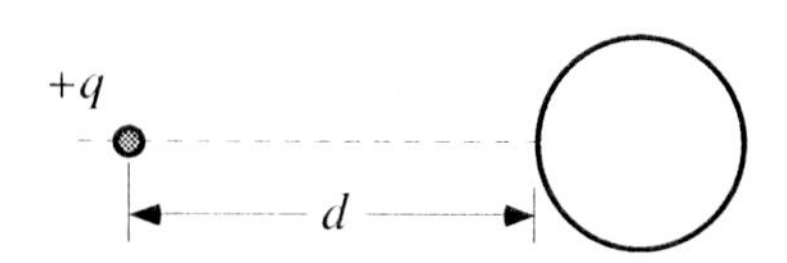

For each change listed, state whether the magnitude of the force exerted on the point charge by the sphere *increases*, *decreases*, or *remains the same*. (Assume that all of the other given variables remain the same for each change given.)

	Effect on the Force exerted on Point Charge			
Change	*No force*	*Increases*	*Decreases*	*Remains the Same*
1. Increase the distance *d*				
2. Increase the sphere diameter keeping the charge a distance *d* away				
3. Increase the positive charge of the point charge				
4. Change the sign of the point charge				
5. Add negative charge to the sphere				

Now for each change listed, state whether the magnitude of the force exerted on the sphere by the point charge *increases*, *decreases*, or *remains the same*. (Assume that all of the other given variables remain the same for each change given.)

	Effect on the Force exerted on the Sphere			
Change	*No force*	*Increases*	*Decreases*	*Remains the Same*
6. Increase the distance *d*				
7. Increase the sphere diameter keeping the charge a distance *d* away				
8. Increase the positive charge of the point charge				
9. Change the sign of the point charge				
10. Add negative charge to the sphere				

ET3-LMCT5: POSITIVE CHARGE IN A UNIFORM ELECTRIC FIELD—ELECTRIC FORCE

A particle with a charge $+q$ is placed in a uniform electric field.

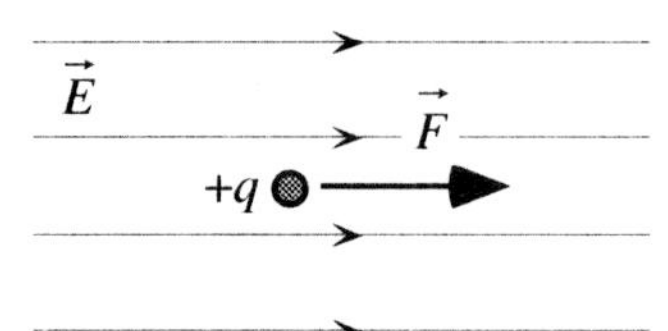

Identify from choices (a)-(f) how each change described in 1 to 7 will impact the electric force on the particle.

This change will:

(a) change only the direction of the electric force.

(b) increase the magnitude of the electric force.

(c) decrease the magnitude of the electric force.

(d) increase the magnitude and change the direction of the electric force.

(e) decrease the magnitude and change the direction of the electric force.

(f) not effect the electric force.

All of these modifications are changes to the initial situation shown in the diagram.

1) **The charge q on the particle is doubled.** ______

2) **The sign of the charge q on the particle is changed to the opposite sign.** ______

3) **The direction of the uniform electric field is rotated 90° clockwise.** ______

4) **The magnitude of the uniform electric field is halved.** ______

5) **The particle is given a push causing a leftward initial velocity.** ______

6) **The direction of the uniform electric field is rotated by 180°.** ______

7) **The particle is given a push causing an upward initial velocity.** ______

ET3-LMCT6: POTENTIAL VS POSITION GRAPH II—FORCE

Shown is a graph of electric potential versus position for a region of space. The labeled points indicate seven locations within this region.

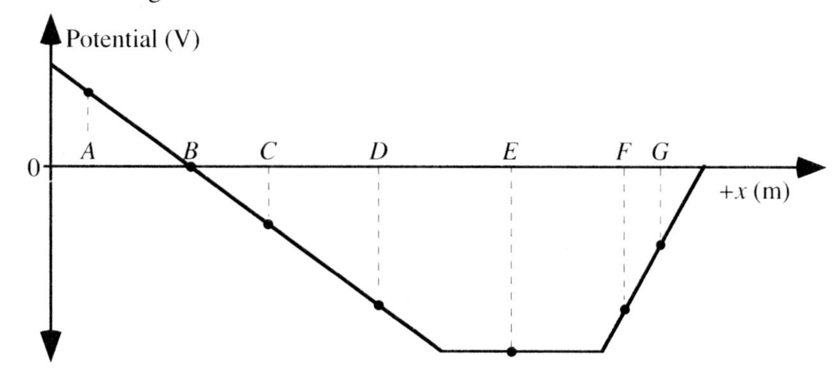

Use one or more of the letters for the labeled locations in answering the questions, or write "none."

1) At which location(s) would a positive charge have the largest force on it? ______

2) At which location(s) would a negative charge have no force on it? ______

3) At which, if any, of the location(s) is the electric field zero? ______

4) At which location(s), if any, is the electric field in the same direction as at *A*? ______

5) At which location(s), if any, is the electric field in the same direction as at *B*? ______

6) At which location(s), if any, does the electric field have the same magnitude field as it does at *C*? ______

ET5-LMCT1: CHARGED INSULATING SHEETS—ELECTRIC FIELD

Two very large, parallel insulating sheets have the charge densities shown. (Only a small portion near the center of the sheets is shown.)

For each change, state whether the magnitude of the electric field at point *P* just to the left of the left sheet *increases, decreases,* or *remains the same.* Each change described is made to the situation in the diagram at right.

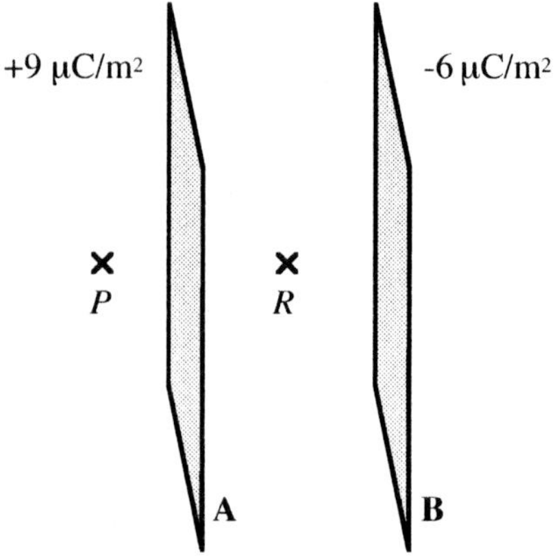

Change	Increases	Decreases	Same
1. Sheet B is moved to the left			
2. Sheet A is moved to the right			
3. Sheet A and sheet B exchange positions			
4. The area of both sheets is doubled			
5. The charge density of sheet A is changed to –3 μC/m^2			
6. The charge density of sheet B is decreased			
7. The sign of the charge on sheet B is changed			
8. A positive point charge is placed at point *R*			
9. Point *P* is moved a small distance to the left			

For each change described, state whether the magnitude of the electric field at point *R* midway between the sheets *increases, decreases,* or *remains the same*. Each change described is made to the situation in the diagram above. Note that when one of the sheets is moved, the point *R* remains fixed at its original location.

Change	Increases	Decreases	Same
10. Sheet B is moved to the left but is still right of point *R*			
11. Sheet A is moved to the right but is still left of point *R*			
12. Sheet A and sheet B exchange positions			
13. The area of both sheets is doubled			
14. The charge density of A is changed to –3 μC/m^2			
15. The charge density of B is changed to –12 μC/m^2			
16. The sign of the charge on sheet B is changed			
17. A positive point charge is placed at point *P*			
18. Point *R* is moved closer to sheet A			

ET7-LMCT1: SIX CHARGES IN THREE DIMENSIONS—FIELD AND POTENTIAL AT ORIGIN

Six point charges are fixed at the same distance away from the origin. All charges are either $+Q$ or $-Q$. In each case, the sign of one of the charges is changed to the opposite sign.

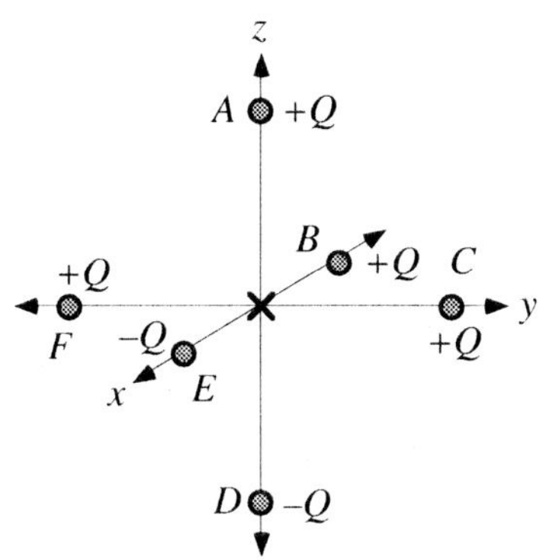

For each change below, state whether (a) the magnitude of the electric field at the origin and (b) the electric potential at the origin *increases*, *decreases*, or *remains the same* as compared to the original case.

Change sign of charge:	(a) Magnitude of the electric field at the origin	(b) Electric potential at the origin
A		
B		
C		
D		
E		
F		

ET7-LMCT2: Four Charges in Two Dimensions—Field and Potential

Four identical point charges are fixed at the same distance from point P. The charges are either $+Q$ or $-Q$.

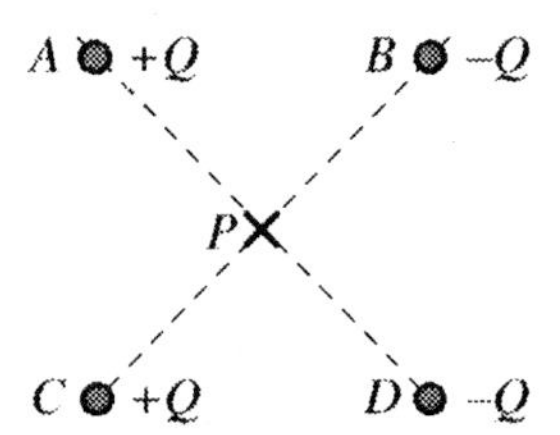

Each action described is made to the situation shown (*i.e.*, "Change sign of charge D" means that charges A, C, and D will be positive and charge B will be negative.)

For each modification:

- **State whether the magnitude of the electric field at the origin *increases, decreases,* or *remains the same.***
- **State whether the electric potential at the origin *increases, decreases,* or *remains the same*** (Use the convention that the electric potential is zero far from the charges.)
- **Indicate the direction of the electric field at the origin after the modification.**

Modification	Electric field	Electric potential	Direction of electric field
1. Change sign of charge *A*			
2. Change sign of charge *B*			
3. Change sign of charge *C*			
4. Change sign of charge *D*			
5. Change signs of charges *B* and *D*			
6. Exchange charges *A* and *B*			
7. Exchange charges *A* and *D*			

ET8-LMCT1: THREE POINT CHARGE SYSTEM—ELECTRIC POTENTIAL

The figure displays three different point charges arranged in a triangle.

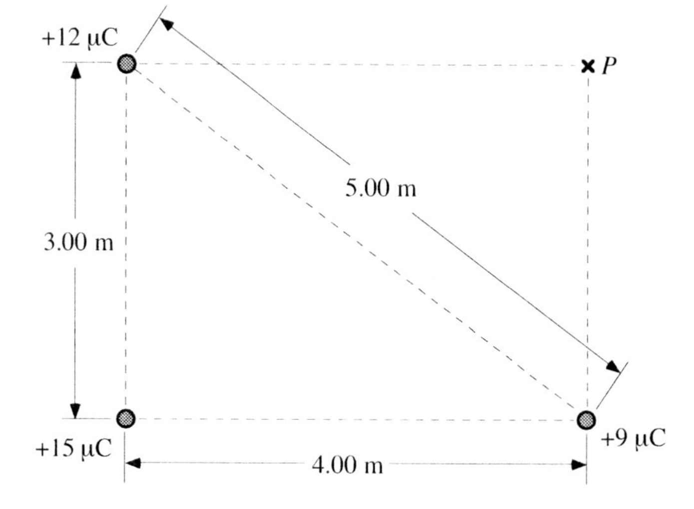

Identify from choices (a)-(d) how each change described in 1 through 7 in turn will impact the original electric potential at point *P*.

This change will:

(a) increase the electric potential at point *P*.

(b) decrease the electric potential at point *P*.

(c) not affect the electric potential at *P*.

Or,

(d) The effect of this change cannot be determined with the given information.

If you choose (d), please explain below.

1) **One of the charges is removed.** ______

2) **The magnitude of one of the charges is increased.** ______

3) **The sign of one of the charges is changed to the opposite sign.** ______

4) **The sign of one of the charges is changed to the opposite sign and its magnitude doubled.** ______

5) **The magnitude of one of the charges is doubled while another is removed.** ______

6) **The distance between two of the charges is decreased.** ______

7) **The signs of all three charges are changed to the opposite sign.** ______

ET10-LMCT1: TWO PARALLEL PLATES—CAPACITANCE

The capacitor in the figure has a cross-sectional area A_o and a distance between plates y. The thickness of each metal plate is t.

A battery with voltage V_o is connected across the plates of the capacitor. There is nothing between the plates of the capacitor.

For each modification listed below, state whether the capacitance *increases, decreases,* or *remains the same.*

Modification	Effect on capacitance		
	Increases	*Decreases*	*Remains Same*
1. Increase the distance y			
2. Increase the area A_o			
3. Increase the plate thickness t			
4. Insert a dielectric material between the plates			
5. Increase the battery voltage V_o			
6. Reverse the battery polarity			

For each change listed below, state whether the magnitude of the charge on the upper plate of the capacitor *increases, decreases,* or *remains the same.*

Modification	Effect on amount of charge on upper plate		
	Increases	*Decreases*	*Remains Same*
7. Increase the distance y			
8. Increase the area A_o			
9. Increase the plate thickness t			
10. Insert a dielectric material between the plates			
11. Increase the battery voltage V_o			
12. Reverse the battery polarity			

CONFLICTING CONTENTION TASKS (CCT)

ET1-CCT1: BREAKING A CHARGED INSULATING BLOCK—CHARGE DENSITY

A block of insulating material has a positive charge distributed uniformly throughout its volume. The block is broken into two unequal pieces, A and B, as shown.

Three students make the following statements about the charge density of pieces A and B:

Alicia: *"Charge density is the charge divided by the volume, and the volume of B is smaller. Since the charge is uniform, and the volume is in the denominator, the charge density is larger for B."*

Bao: *"The charge density of piece A is larger than the charge density of piece B. Piece A is larger, so it has more charge."*

Craig: *"They both have the same charge density. It's still the same material."*

Which of these students is correct?

Alicia_____ Bao_____ Craig_____ None of them______ **Explain.**

ET1-CCT2: CHARGED INSULATORS CONNECTED WITH A SWITCH—CHARGE

Two solid, insulating spheres are connected by a wire and a switch. The spheres are the same size but they have different initial uniform charge distributions.

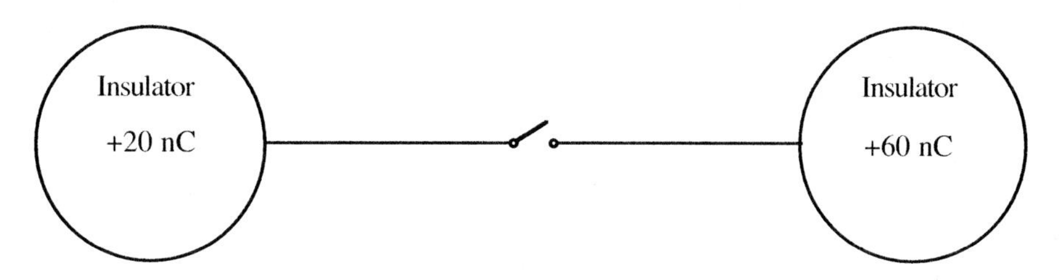

Three students are discussing what would happen if the switch was closed.

Arturo: *"Since the spheres are the same size, charge will move until there is an equal charge of 40 nC on each."*

Beth: *"I agree, but since they are insulators, the charge will move very slowly. Eventually there will be the same charge of 40 nC on each, but it will take a long time, perhaps 5-10 minutes.*

Caitlin: *"No, since they are insulators the charge cannot move. It doesn't matter whether the switch is open or closed."*

Which of these students is correct?

Arturo_____ Beth_____ Caitlin_____ None of them_____ **Explain.**

ET1-CCT3: CHARGED SHEET—ENCLOSED CHARGE

A plastic sheet has a uniform surface charge density σ. An imaginary (Gaussian) cylinder has its left and right end caps equal distances from the sheet, and has a cross-sectional area A.

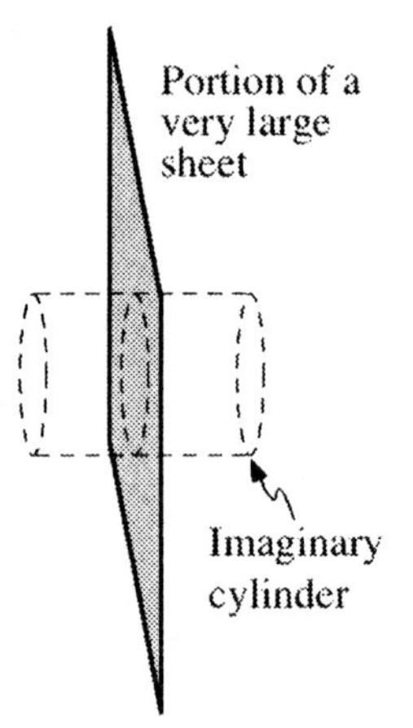

Three students make the following statements about the charge enclosed by the cylinder:

Abby: *"The charge enclosed is just σ. Since the sheet is uniform, it doesn't make any difference how big the imaginary surface is."*

Barack: *"There is no charge enclosed. The field through the left end cap points to the left and the field through the right end cap points to the right. These fields cancel, and the net electric flux is zero. So the charge enclosed is zero by Gauss' law."*

Carlos: *"To find the charge enclosed we would need to know something about the length of the cylinder, which isn't given."*

Which of these students is correct?

Abby_____ Barack_____ Carlos_____ None of them_____ **Explain.**

ET3-CCT1: ELECTRON IN A UNIFORM ELECTRIC FIELD—ELECTRIC FORCE

Consider the following statements about the motion of an electron placed at rest in a uniform electric field as shown:

Anna: *"Since the electron is negative, it will move downward at a constant velocity proportional to the strength of the electric field."*

Brooke: *"The electron will accelerate upward because particles move in the direction of the electric field, which points upward."*

Chico: *"The electron will move downward because it is a negative particle. The force acting on it will be opposite the direction of the electric field. It will move with a constant acceleration."*

Which of these students is correct?

Anna_____ Brooke_____ Chico_____ None of them_____ **Explain.**

ET3-CCT2: TWO CHARGES—FORCE

Two negatively charged particles labeled A and B are separated by a distance x. The particles have different charges and masses as shown.

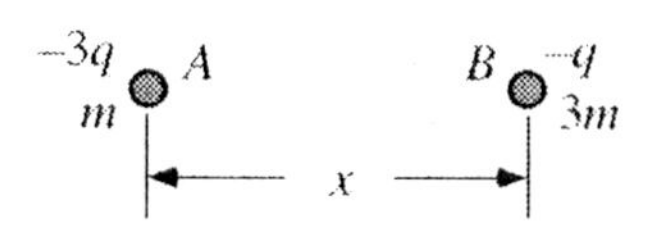

Three students are discussing what will happen just after the two particles are released.

Antonio: *"The magnitude of the force that A exerts on B will be the same as the magnitude of the force that B exerts on A. Since A has less mass, it will have a larger acceleration."*

Brenda: *"The magnitude of the force on A by B is greater than the magnitude of the force on B by A since B has more mass. So A will have the largest acceleration."*

Cho: *"A has more charge but it has less mass. The larger mass of B is exactly compensated for by the larger charge of A. The acceleration of both will be the same."*

Which of these students is correct?

Antonio_____ Brenda_____ Cho_____ None of them_____ **Explain.**

ET3-CCT3: SPHERE AND A POINT CHARGE—FORCE

In each case shown at right there is a point charge $+q$ a distance d from the closest point of a neutral metal sphere. The sphere in case B has a larger diameter than the sphere in case A. Three students are comparing the two cases:

Case A

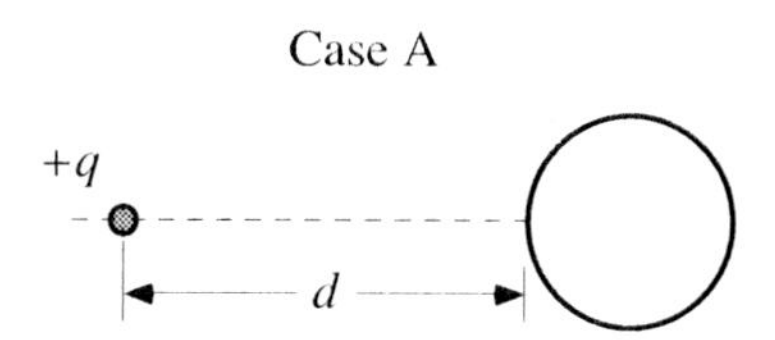

Case B

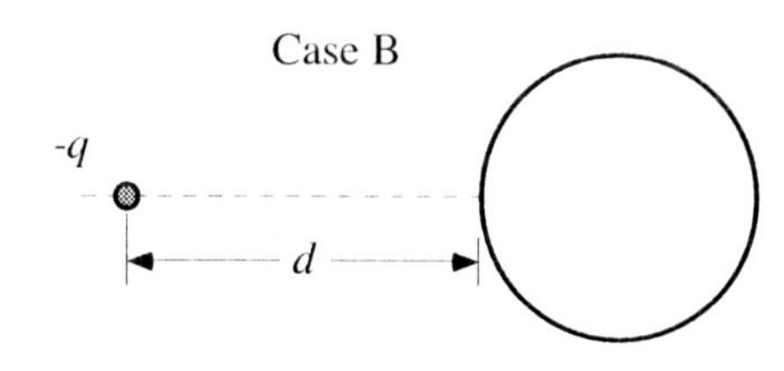

Aaron: *"I don't think there would be any electric forces in either case. Since the sphere has no net charge, there is no attraction or repulsion."*

Bao: *"The forces on the point charges are equal in the two cases. There is an attraction because the point charge will pull the electrons in the sphere toward it. But the distance between the point charge and the electrons is the same in both cases, so the force of attraction is the same."*

Carlota: *"When the electrons are pulled toward the point charge, they leave a pool of positive charges on the other side of the sphere. These positive charges repel the point charge, and this balances the attraction of the electron. The sphere overall is still neutral, so there is as much positive charge as negative charge, and there is no net force between the objects."*

Which of these students is correct?

Aaron_____ Bao_____ Carlota_____ None of them_____ **Explain.**

ET3-CCT4: CURVED CHARGED ROD AND TWO POINT CHARGES—FORCE

A charge $+q$ lies a distance d from a charge $+Q_o$. An insulating curved charged rod with charge $+Q_o$ uniformly distributed is positioned as shown. All points of the curved rod are also a distance d from $+q$. Consider the following statements made by three students:

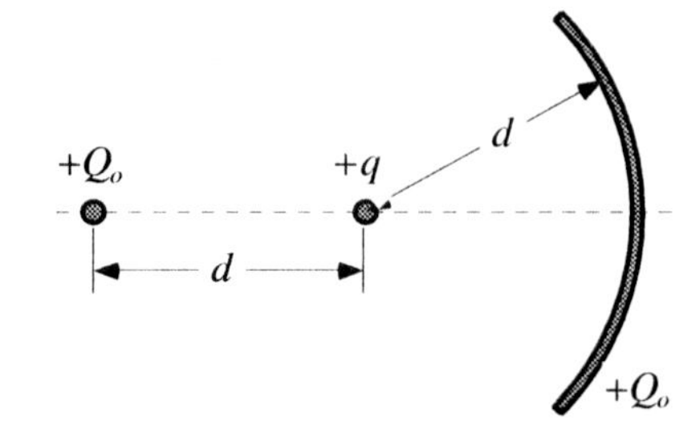

Anita: *"The charges that are exerting forces on $+q$ are both the same distance away, and the same charge. So from Coulomb's law, the forces have the same magnitude. Since they are in opposite directions, the net force on $+q$ is zero."*

Banji: *"The net electric force on $+q$ is to the right. The test charge is perpendicular to the middle of the rod. Therefore, the force from the bottom half of the rod cancels the force from the top half, and you are left with a repellant horizontal force from the charge on the left."*

Christina: *"It is only the vertical components of the forces due to the charge on the rod that cancel out. The horizontal components all act to produce a force on $+q$ to the left. So, there is still a charge $+Q_o$ producing a horizontal force to the left, just as there is a charge $+Q_o$ to the left of $+q$ producing a force to the right. The net force on $+q$ is zero."*

Which of these students is correct?

Anita_____ Banji_____ Christina_____ None of them_____ **Explain.**

ET3-CCT5: PAIRS OF CHARGED CONDUCTORS—FORCE

Two conducting spheres have the same radius but have different charges. Three students are discussing whether Coulomb's law applies when calculating the force one sphere exerts on the other.

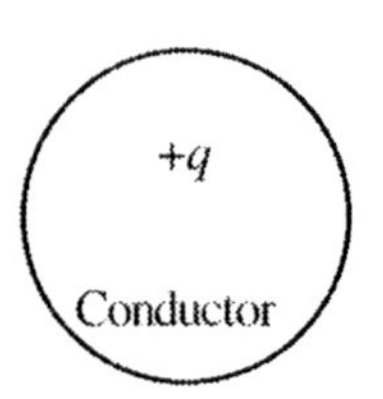

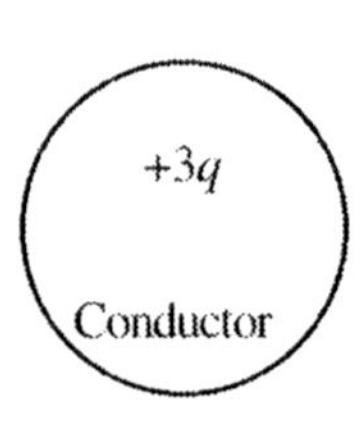

Alejandro: *"Since these are spheres, they have the same symmetry as points, and Coulomb's law applies."*

Belinda: *"I agree, but only because these are positive charges. If they were negative charges they would be free to move within the spheres, and the distance would change."*

Colin: *"I disagree. Coulomb's Law only applies to point charges. Since the conducting spheres are not points, it cannot be used."*

Which of these students is correct?

Alejandro_____ Belinda_____ Colin_____ None of them_____ **Explain.**

ET3-CCT6: CONDUCTING CUBE BETWEEN POINT CHARGES—FORCE

In case A, two equal and opposite charges are separated by a distance *d*. Case B is identical to case A except that a neutral metal cube has been placed between the two charges. Four students are comparing the electric force on the positive charge in the two cases:

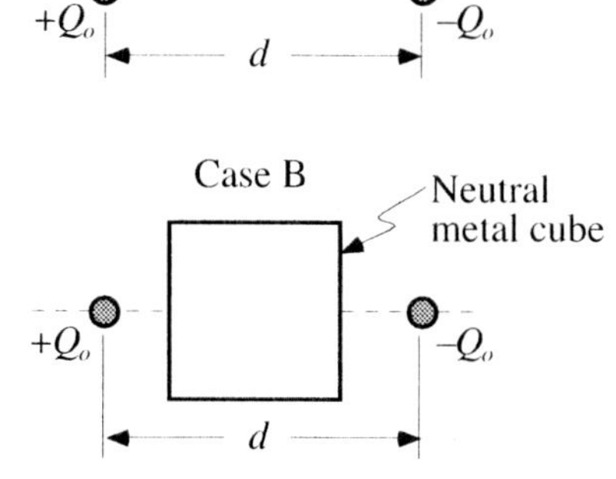

Alicia: *"Since the block is a conductor, it lets more charge travel between the point charges. The force will be stronger."*

Boris: *"The electric field inside of a conductor is zero. So the metal cube blocks the electric force on the positive charge by the negative charge. There might be some field lines that still attract the positive charge by going around the metal cube, but the force is much smaller in case B."*

Cody: *"If the cube is a perfect conductor, they will be equal, since then the cube would not interfere at all with the charge. Otherwise the force would be greater in case A."*

Delia: *"B is greater than A. The cube is a conductor! It is as if the distance in the cube 'wasn't there' because of the permittivity constant of the metal cube."*

Which of these students is correct?

Alicia_____ Boris_____ Cody_____ Delia_____ None of them_____ **Explain.**

ET4-CCT1: CART APPROACHING SPHERE—DISTANCE

An electrically charged sphere mounted on a cart is approaching a fixed electrically charged sphere. At the instant shown, the spheres are one meter apart and the cart has a kinetic energy of 20 Joules. Ignore friction and the charges on the spheres are given in the figure. Three students considering this situation make the following contentions:

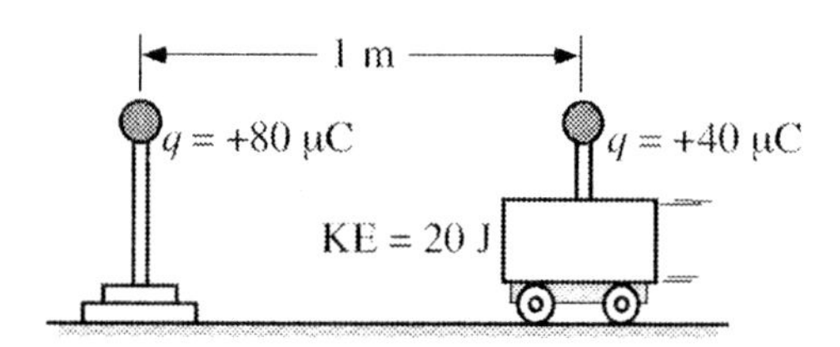

Alan: *"The cart will slow down as it approaches the fixed sphere, stop and reverse its motion. We cannot figure out how close the cart will get because we don't know the mass so we can't find the acceleration."*

Brad: *"While I agree that the cart will slow down and stop, I don't think you can figure out how close it will come to the fixed sphere because the force will be changing."*

Carlos: *"Well you guys do have the motion right, but you are both wrong about finding how close the cart gets to the sphere. I think you can use energy to find the final distance between spheres."*

Which of these students is correct?

Alan _______ Brad _______ Carlos _______None of these _______ **Explain.**

ET6-CCT1: ELECTRIC FORCE ON A PROTON—ELECTRIC FIELD

The graph shows the electric force in the x-direction acting on a proton at different times.

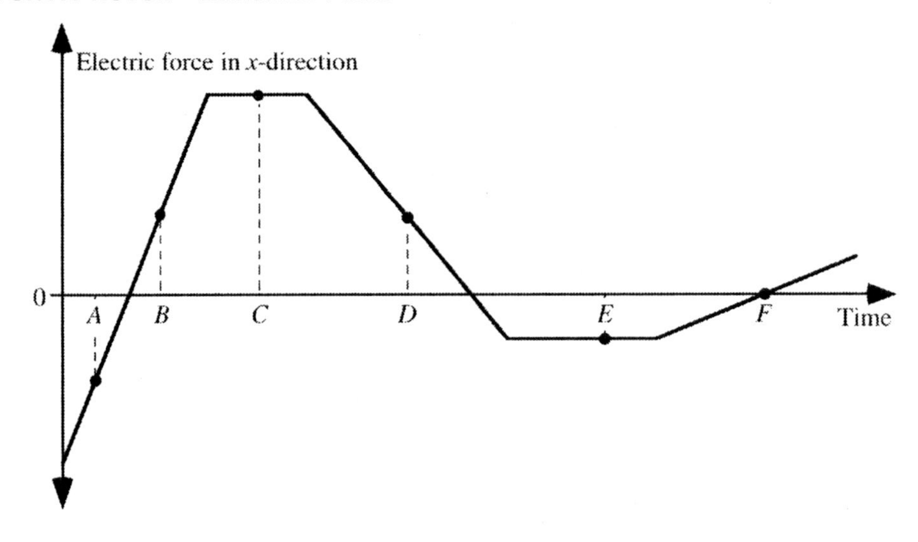

Four students are discussing inferences that might be drawn from this graph:

Amanda: *"The electric fields at C and E are zero because the slope of the line is zero."*

Ben: *"No, the electric potentials at C and E are zero because the slope of the line is zero."*

Caleb: *"I think the electric field is zero at F because the electric force is zero."*

Dakota: *"No, you are all wrong. Force and electric field vary as $1/r^2$, so we should be looking at curved lines, not straight lines."*

Which of these students is correct?

Amanda ______ Ben ______Caleb ______ Dakota ______ None of them ______ **Explain.**

ET6-CCT2: ELECTRIC POTENTIAL VS DISTANCE GRAPH II—ELECTRIC FIELD

Shown is a graph of the electric potential versus distance in a region in which there may be an electric field. The labeled points indicate six locations within this region.

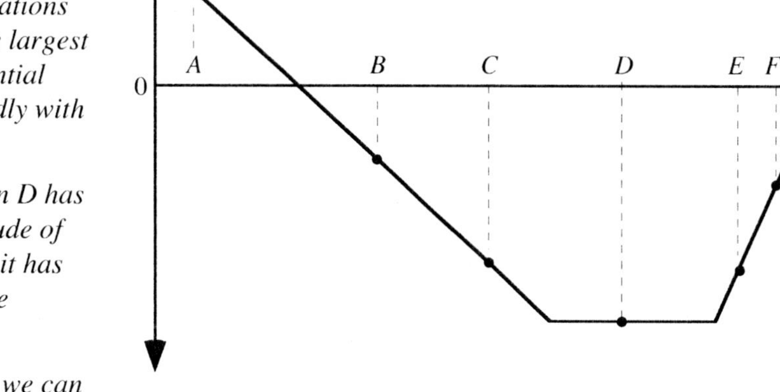

Three students considering this graph make the following contentions.

Angel: *"The magnitude of the electric field at locations E and F will be the largest because their potential changes most rapidly with distance."*

Bo: *"No, I think location D has the largest magnitude of electric field since it has the largest negative potential."*

Cathy: *"Well I don't think we can tell anything about the electric fields since potential is a scalar, but electric field is a vector."*

Which of these students is correct?

Angel______ Bo______ Cathy______ None of them______ **Explain.**

ET5-CCT3: THREE-DIMENSIONAL LOCATIONS IN A CONSTANT ELECTRIC POTENTIAL—FIELD

The electric potential has a constant value of 12 volts everywhere in a three-dimensional region, part of which is shown.

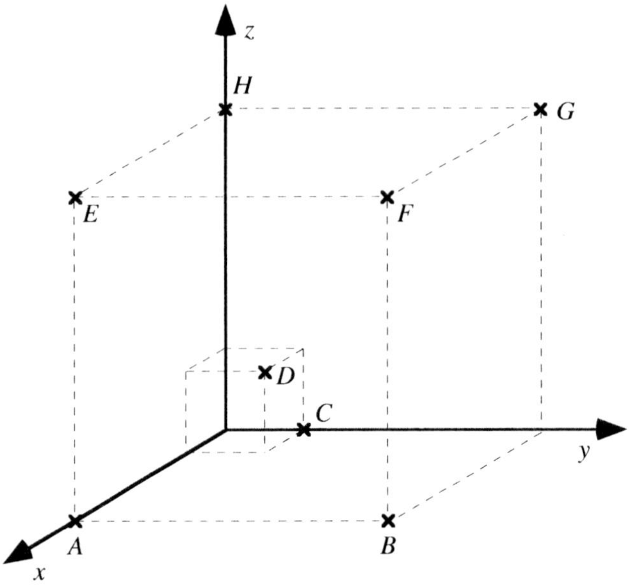

Consider the following statements made by students discussing the electric field strength (magnitude) at the labeled points:

Aliyah: *"I think the electric field strength just depends on the distance from the origin. So point C is largest since it is closest to the origin and D is a little smaller. The smallest electric field strength is at F since it is the largest distance away from the origin."*

Brian: *"I do not think you can determine the relative electric field strengths at these points, since we don't know how fast the electric field is changing."*

Chao: *"I think they are all zero because the electric potential is constant."*

Diane: *"I think they are all the same but not zero since there is a constant potential."*

Which of these students is correct?

Aliyah ______ Brian______ Chao ______ Diane ______ None of them ______ **Explain.**

ET5-CCT4: THREE CHARGES IN A LINE—ELECTRIC FIELD

Shown are two cases where three charges are placed in a row.

Three students are comparing the electric field that exerts a force on the middle charge in the diagrams.

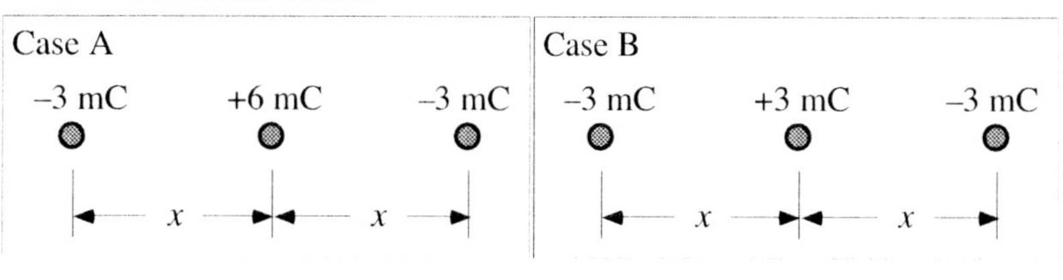

Adrianna: *"All three charges contribute by the principle of superposition. So the field is going to be greatest in case A since the contributions due to the three charges will be greatest."*

Brandon: *"I think it's a bogus question. The field at that point is undefined because there is a charge there."*

Catalina: *"You're wrong. The field that exerts a force on the middle charge is the field due to all of the other charges in the region. Since these don't change, the field acting on the middle charge is the same in both cases."*

Which of these students is correct?

Adrianna ______ Brandon ______ Catalina______ None of them______**Explain.**

ET5-CCT5: AIRPLANE FLYING BETWEEN TWO CHARGED CLOUDS—ELECTRIC FIELD

An airplane is flying at an altitude of 3000 m between two clouds. One cloud at altitude 5000 m has a charge of +30 C and another cloud at altitude 1000 m has a charge of –30 C.

Three students are discussing the electric field at the location of the plane.

+30 C
2000 m
2000 m
-30 C

Ahmed: *"I think the electric field is zero because the plane is the same distance from equal positive and negative charges, and they produce electric fields which cancel out."*

Brianna: *"I think the electric field is zero because the plane is equal distance between equal positive and negative charges which produce an electric potential of zero at that point."*

Coen: *"I think the electric field points downward because the positive and negative charges produce electric fields in the same direction which add together."*

Which of these students is correct?

Ahmed_____ Brianna_____ Coen_____ None of them______ **Explain.**

ET5-CCT6: TWO CHARGED SPHERES—ELECTRIC FIELD

In each case shown at right, the solid sphere has radius a and total charge $+Q$. However, the sphere in case A is an insulator with its charge uniformly distributed throughout its volume, and the sphere in case B is a conductor. Consider the statements below about the electric field at points P and S in the two cases.

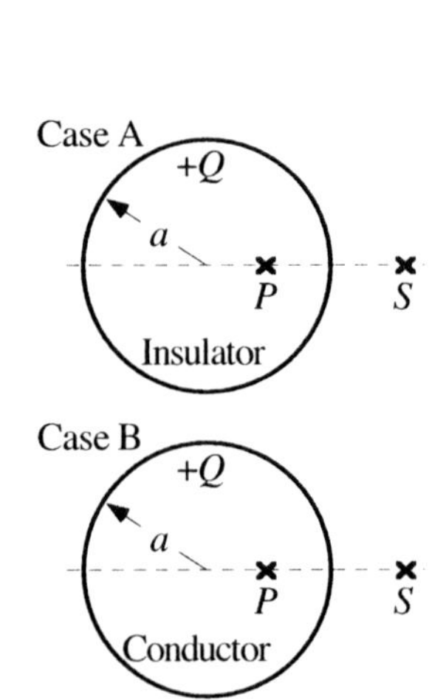

Amber: *"The electric field at point S will be the same for the two cases because the charges of the spheres are the same. Likewise, the electric field at P will be the same."*

Blanca: *"The electric field at S will be different in the two cases because one is a conductor while the other is an insulator. The electric field at P will be the same because they have the same charge. "*

Cecilia: *"The electric field at both points S and P will be different because one is a conductor while the other is an insulator."*

Which of these students is correct?

Amber____ Blanca____ Cecilia____ None of these students____ **Explain.**

ET5-CCT7: POTENTIAL NEAR CHARGES—ELECTRIC FIELD

Three students are considering the situations shown where the electric potential is given at a point that is midway between equal magnitude electric charges. V_0 is the potential due to a single charge at the same distance as shown on the identical grids.

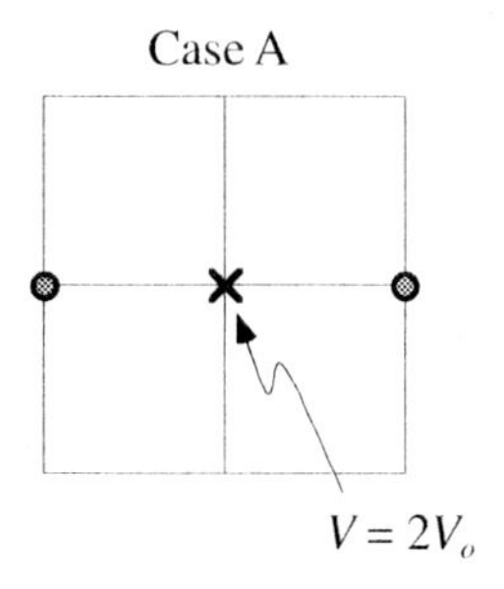

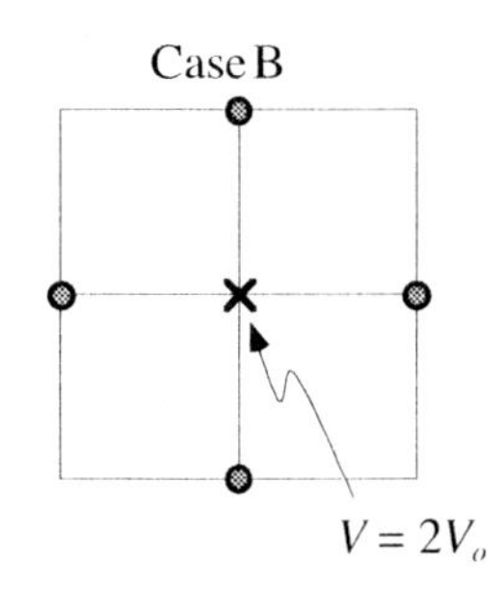

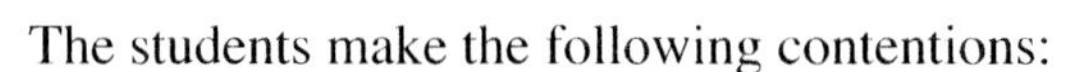

The students make the following contentions:

Alberto: *"The electric fields at the two midpoints will be different because more charge means a bigger field."*

Bernardo: *"I disagree. Adding more charges doesn't always increase the field if the charges have different signs."*

Carol: *"Since the potential hasn't changed with the two extra charges the fields due to those charges must cancel."*

Which of these students is correct?

Alberto_____ Bernardo_____ Carol_____ None of them_____ **Explain.**

ET5-CCT8: POINT CHARGE IN A CONDUCTING SHELL—ELECTRIC FIELD

A point charge is enclosed by a neutral metal shell, but is offset from the center of the shell. Four points are labeled in the diagram. Points *B* and *C* are the same distance from the center of the shell, and points *A* and *D* are the same distance from the center of the shell.

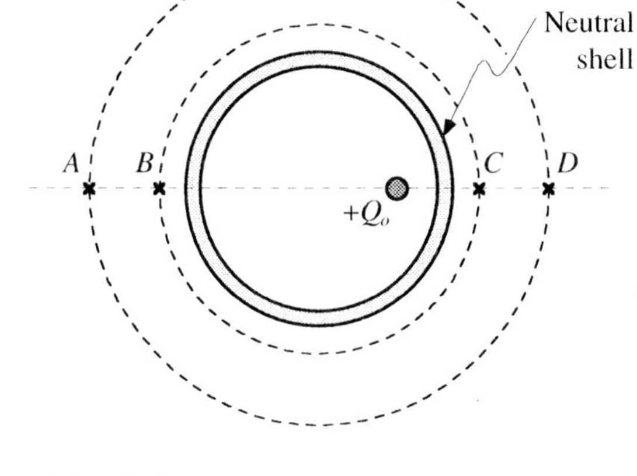

Three students discuss the electric field at the labeled points.

Alfredo: *"The electric field is zero at all of these points. There is no electric field inside a conductor, so the field from the point charge won't get out of the shell. Since there are no other charges nearby, the field is zero."*

Brian: *"I think the electric field is the same at points C and B, and it is weaker but the same at points A and D. The shell is going to become polarized by the point charge, and it will produce a field outside the shell. The total field outside will look just like the field due to a point charge at the center of the shell."*

Carmen: *"I think the field is blocked by the shell but some of it gets through. The closest points to the charge will have a stronger field. The field will be strongest at point C and weakest at point A."*

Which of these students is correct?

Alfredo ______Brian ______ Carmen ______None of them ______ **Explain.**

ET5-CCT9: FIELD OUTSIDE A SPHERE WITH A CAVITY—ELECTRIC FIELD

Shown at right is a cross-section of a sphere made of an insulator (such as plastic) that has a spherical cavity in it. Fixed in position in the cavity is a positive charge $+q$. Points P and S are both one cm away from the surface of the sphere. Three students considering this situation make the following statements:

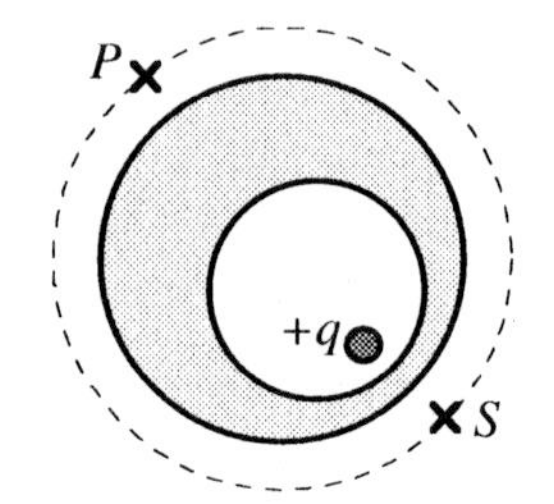

Alberto: *"I think the electric field at the two points will be zero since the sphere is made of an insulating material which will block the electric field of the charge."*

Beatrice: *"If it was a conductor it would be zero. Here, the electric at S has to be larger than at P since the charge inside the sphere is closer to S."*

Cecilia: *"Neither of you are right. Since the sphere is an insulator it will be polarized throughout. There will be an electric field outside of the sphere, but it will be the same everywhere around the sphere."*

Which of these students is correct?

Alberto ______ Beatrice ______ Cecilia ______ None of them _______ **Explain.**

ET6-CCT3: SYSTEMS OF POINT CHARGES—WORK TO ASSEMBLE

Shown below are three arrangements of eight identical positive point charges on identical grids.

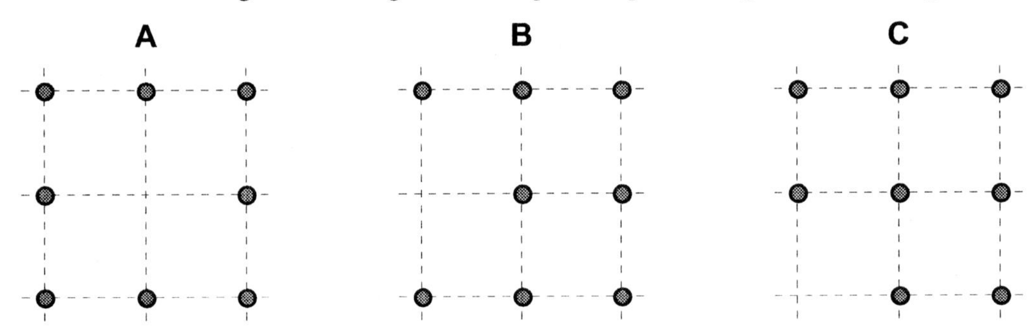

Consider the following statements made by three students:

Andrew: *"Arrangement A requires the least amount of work to assemble because the charges are spread out while B and C require the same amount of work to assemble because the missing charge is on the edge."*

Bonita: *"Arrangements A, B, and C all require the same amount of work to assemble because they all have eight charges."*

Craig: *"You can make arrangement B from A by moving only one charge, and arrangement C from B by moving only one charge. In both cases you have to fight the repulsive force of the other charges, so C requires the most work, then B, then A."*

Which of these students is correct?

Andrew ______ Bonita ______ Craig ______ None of them ______ **Explain.**

ET8-CCT1: Two Charged Spheres—Electric Potential

In each case shown at right, the solid sphere has radius a and a total charge $+Q$. However, the sphere in case A is an insulator with its charge uniformly distributed throughout its volume, and the sphere in case B is a conductor. Consider the statements below about the electric potential at points P and S in the two cases.

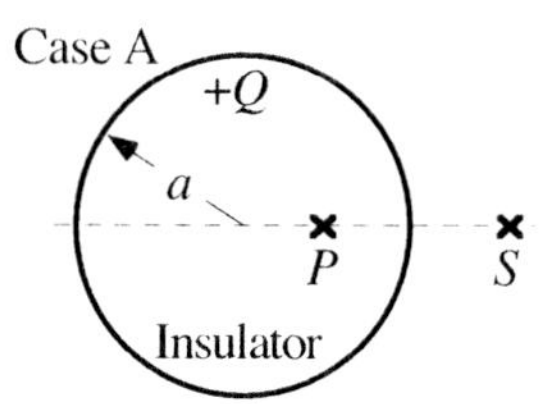

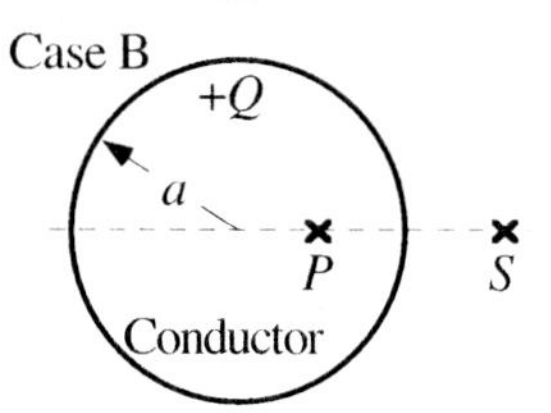

Alex: *"The electric potential at point S will be the same for the two cases because the charges of the spheres are the same. Likewise, the electric potential at P will be the same."*

Brooke: *"The electric potential at S will be different in the two cases because one is a conductor while the other is an insulator. The electric potential at P will be the same because they have the same charge."*

Christina: *"The electric potential at both points S and P will be different because one is a conductor while the other is an insulator."*

Which of these students is correct?

Alex ______ Brooke ______ Christina ______ None of these students _____ **Explain.**

ET9-CCT1: Gaussian Cube near a Charge—Electric Flux

A Gaussian cube is shown to the left of a positively charged particle. Consider the statements below:

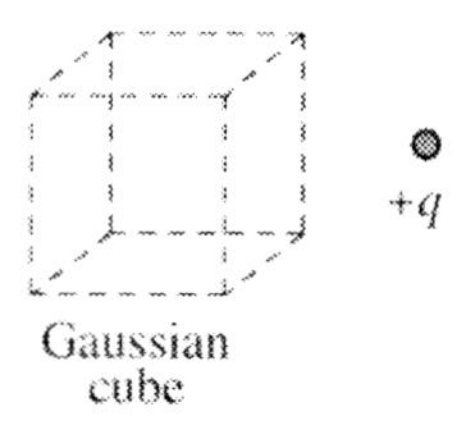

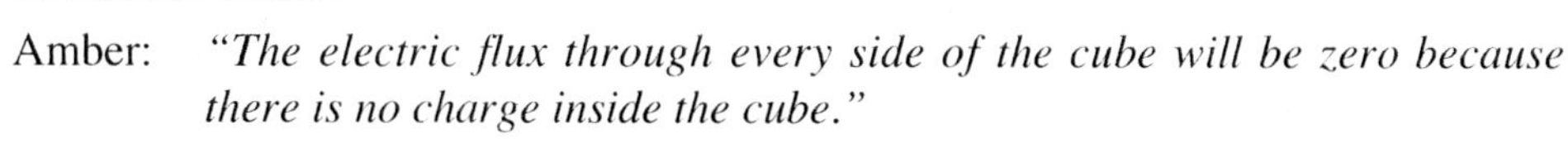

Amber: *"The electric flux through every side of the cube will be zero because there is no charge inside the cube."*

Bo: *"Since there is no charge inside the cube, the net electric flux through the cube will be zero. However, each side will have non-zero electric flux values which, when added together, cancel out."*

Caleb: *"The net electric flux on the cube will not be zero. The number of electric field lines on the side of the cube closest to the particle will be greater than the lines through the opposite side. It will therefore be impossible for the electric flux through each side to cancel."*

Which of these students is correct?

Amber _____ Bo _____ Caleb _____ None of them _____ **Explain.**

ET9-CCT2: CHARGES INSIDE GAUSSIAN SPHERE—ELECTRIC FLUX AND ELECTRIC FIELD

Two charged particles are inside a Gaussian sphere. Consider the statements below:

+5 μC −5 μC

Gaussian sphere

Andrew: *"Since the charges inside the sphere cancel or neutralize each other, the net electric flux through the sphere will be zero, so there will be no electric field."*

Banji: *"Since the net charge inside the sphere is zero, the net electric flux through the sphere will be zero. However, the electric field must have a non-zero value inside and outside of the sphere because charged particles produce electric fields."*

Chris: *"Since there are charges, there must be an electric field present. Since electric flux depends on the electric field and the area of the surface, neither of which is zero, there must be a non-zero net electric flux through the sphere."*

Which of these students is correct?

Andrew_____ Banji_____ Chris_____ None of them_____ **Explain.**

ET10-CCT1: CHARGED ROD NEAR A SUSPENDED BAR MAGNET—ROTATION

A bar magnet is suspended by a string. With the magnet held in place, a charged rod is brought close to the point of suspension of the magnet as shown. The suspended bar magnet is then released so that it is free to rotate.

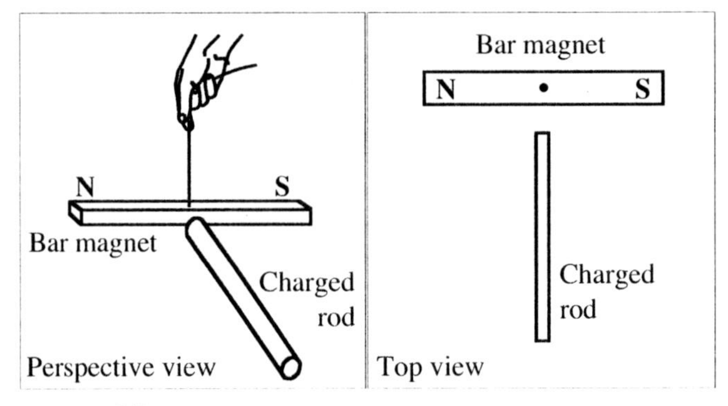

Three students are discussing what will happen when the magnet is free to rotate.

Aaron: *"I think the magnet will rotate clockwise when viewed from above. The north pole will be repelled by the positive charge, and the south pole will be attracted."*

Ben: *"I think the magnet will just sit there. A north pole is not like a positive charge."*

Carl: *"It's more complicated than that. The magnet will induce a north pole on the charged rod because the magnetic field will cause the electron spins to align. Since there are then two north poles close to each other, the magnet will rotate clockwise."*

Which of these students is correct?

Aaron_____ Ben_____ Carl_____ None of them_____ **Explain.**

ET10-CCT2: Charged Rod and Electroscope—Deflection

A positively charged rod is brought near an electroscope. Even though the rod does not touch the electroscope, the leaf of the electroscope deflects. Below, three students discuss this demonstration.

Alfredo: *"There are positive charges that jump from the rod to the plate of the electroscope. Since the electroscope is now charged, the leaf moves out."*

Brent: *"No charges move from the rod to the plate. When the rod comes close, electrons in the electroscope move toward the plate. This leaves the bottom of the electroscope positively charged, and the leaf lifts."*

Carmen: *"Positive charges are fixed in place. When the rod is brought close to the electroscope plate, the electrons in the plate are attracted and jump to the rod. This leaves the electroscope positively charged, and the leaf lifts."*

Which of these students is correct?

Alfredo_____ Brent_____ Carmen_____ None of them_____ **Explain.**

CHANGING REPRESENTATION TASKS (CRT)

ET5-CRT1: ELECTRIC FORCE ON AN ELECTRON—ELECTRIC FIELD

The graph below shows the electric force acting on an electron at different times.

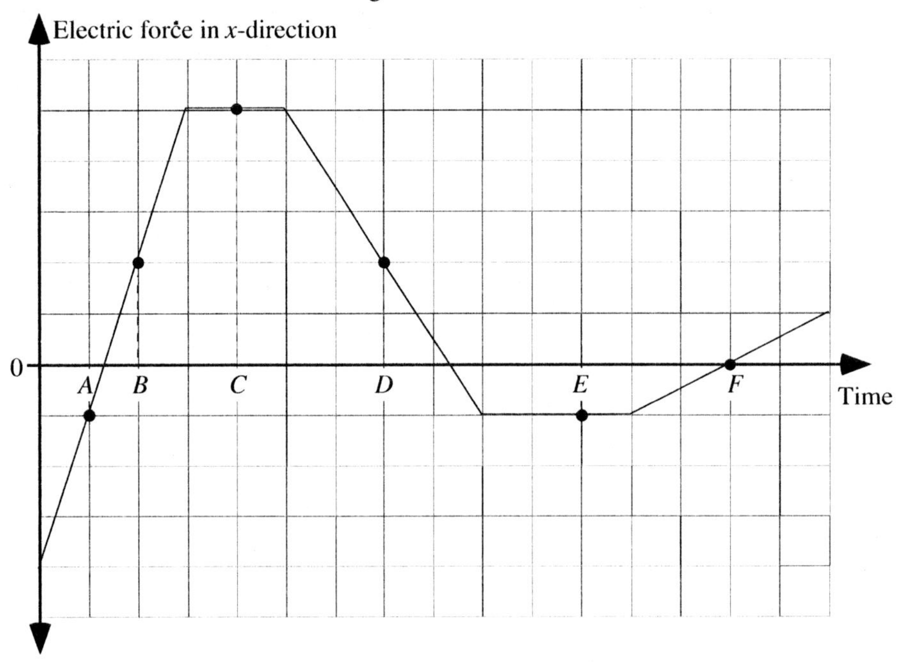

Sketch below the corresponding graph for the electric field.

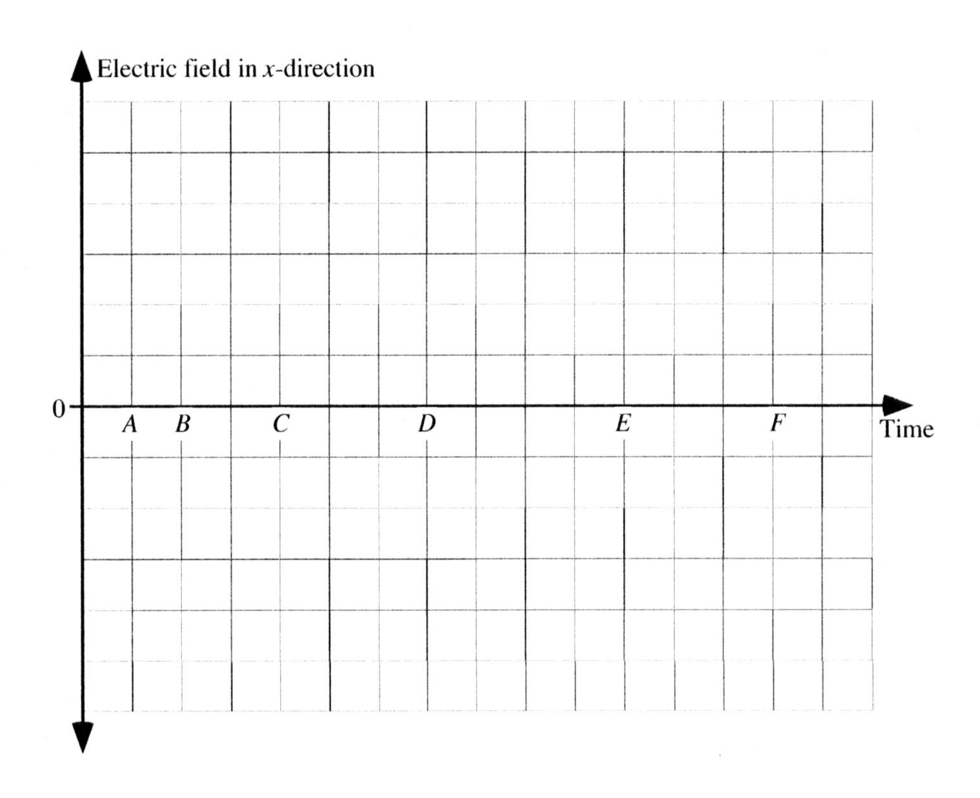

ET3-CRT1: CHARGES AND EQUIPOTENTIALS—FORCE

The dashed lines in the figure below show equipotentials in a region in which there is an electric field. A positive point charge is placed at each of the seven labeled points in turn.

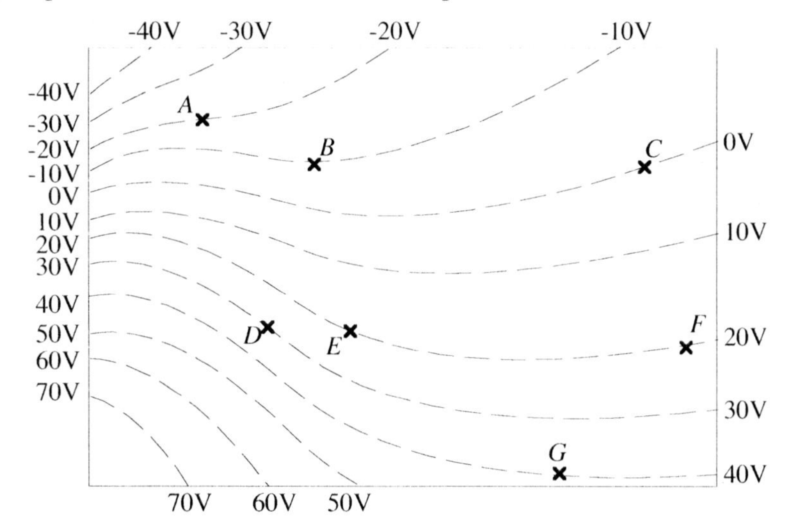

In the diagram above, draw arrows to show the direction of the force on the charge at each of the labeled points. Explain.

ET5-CRT2: POTENTIAL VS POSITION GRAPH II—ELECTRIC FIELD DIRECTION

Shown below is a graph of potential versus position for a region of space. The labeled points indicate seven locations within this region.

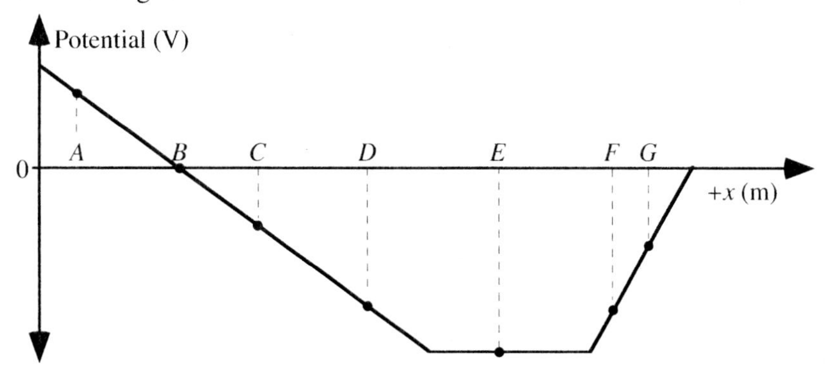

The arrow in the leftmost box represents the electric field direction at point *A*.

Draw arrows in the other boxes to represent the direction of the x-component of the electric field at points *B* through *G*. If there is no field at a labeled point, state that explicitly.

at *A*	at *B*	at *C*	at *D*	at *E*	at *F*	at *G*
→						

Explain.

ET5-CRT3: POTENTIAL VS POSITION GRAPH—ELECTRIC FIELD GRAPH

A graph of the electric potential in a region of space as a function of x is shown below.

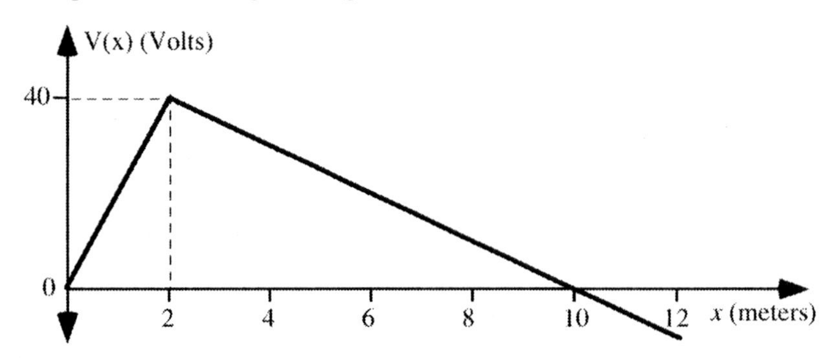

Graph the x-component of the corresponding electric field. Include values on the vertical axis. **Explain.**

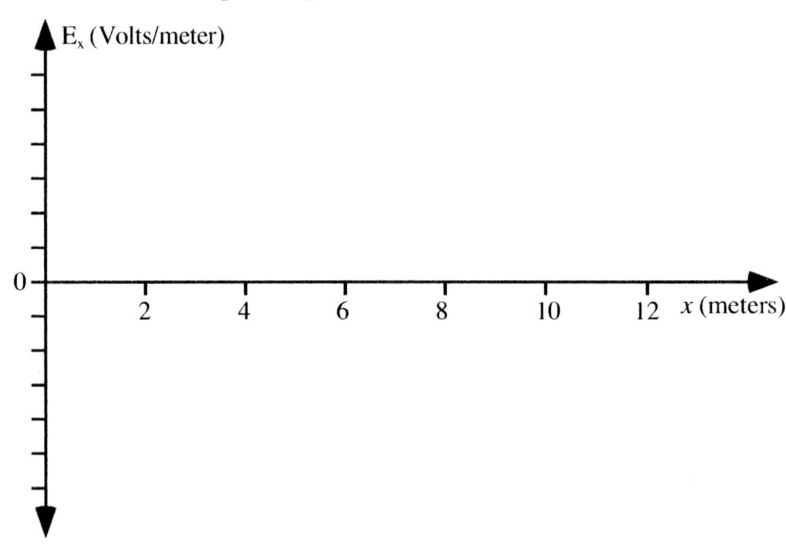

ET8-CRT1: PARALLEL PLATE CAPACITOR—GRAPH OF POTENTIAL I

A 30-volt battery is connected across the plates of the capacitor. Assume the plates are large enough that the field is uniform in the region of interest. Use a coordinate system where $+x$ is in the horizontal direction to the right and $+y$ is in the vertical direction upward on the page. Assume that $V = 0$ at the lower plate.

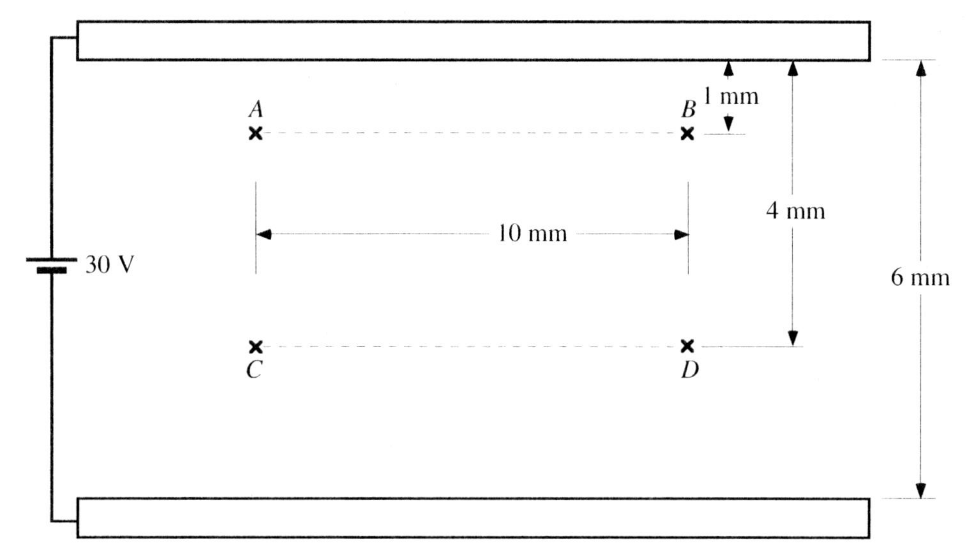

(1) Graph the electric potential vs. *x* for the line from *A* to *B*. Be sure to include values. **Explain.**

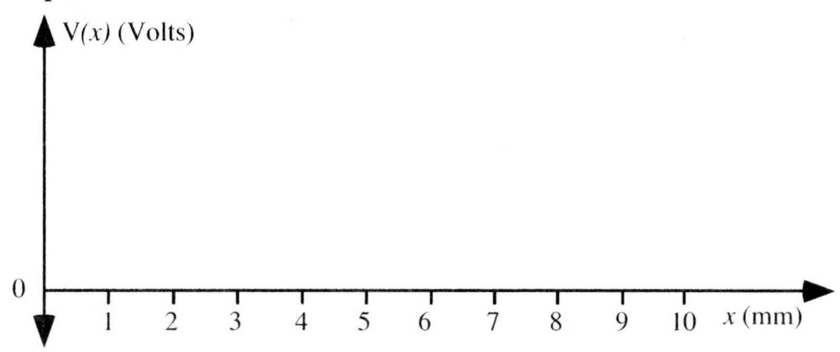

(2) Graph the electric potential vs. *x* for the line from *C* to *D*. Be sure to include values. **Explain.**

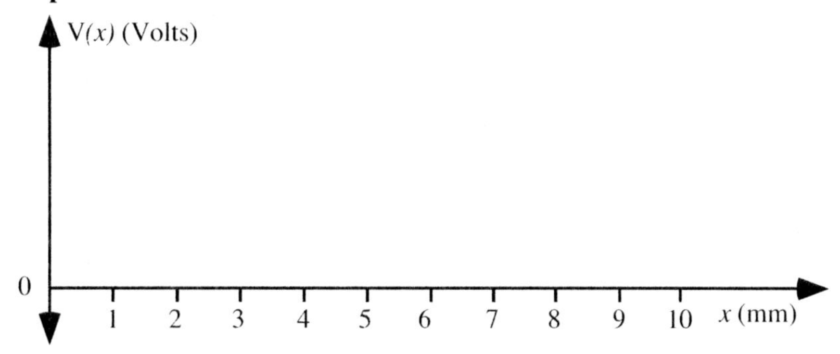

ET10-CRT1: PARALLEL PLATE CAPACITOR—GRAPH OF POTENTIAL II

The capacitor in the figure below has a cross-sectional area *A*, and a distance between plates of 6 mm. Assume the plates are large enough that the field is uniform in the region of interest. A 30-volt battery is connected across the plates of the capacitor. There is nothing between the plates of the capacitor. The dotted line *AB* is located along the center of the capacitor, and the dotted line *CD* is 3 mm to the right of *AB*. Use a coordinate system where $+x$ is in the horizontal direction to the right and $+y$ is in the vertical direction upward on the page. Assume that $V = 0$ at the lower plate.

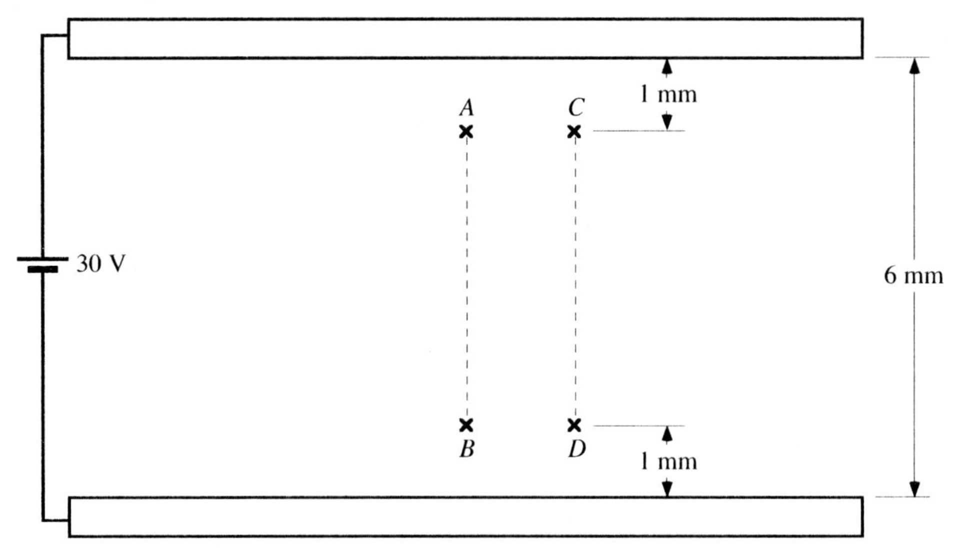

(1) Graph the electric potential vs. *y* for the line from *A* to *B*. Be sure to include values. **Explain.**

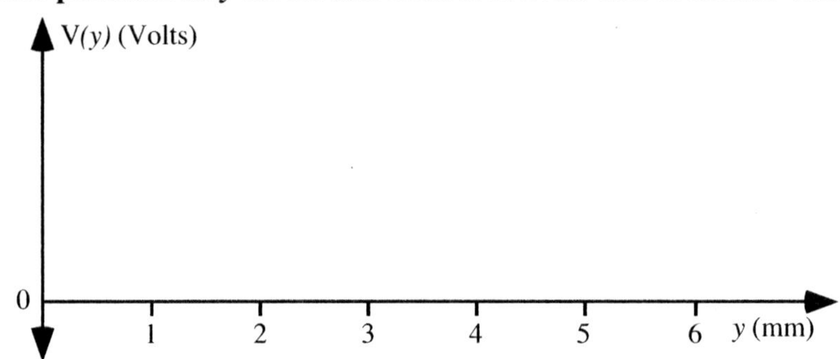

(2) Graph the electric potential vs. *y* for the line from *C* to *D*. Be sure to include values. **Explain.**

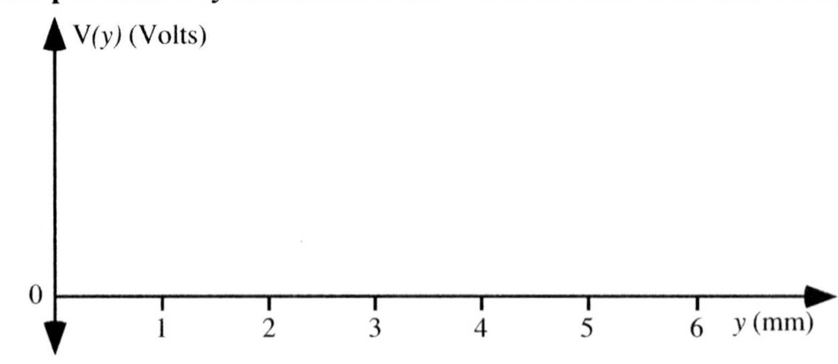

Bar Chart Tasks (BCT)

ET1-BCT1: Charged Insulating Blocks—Charge and Charge Density

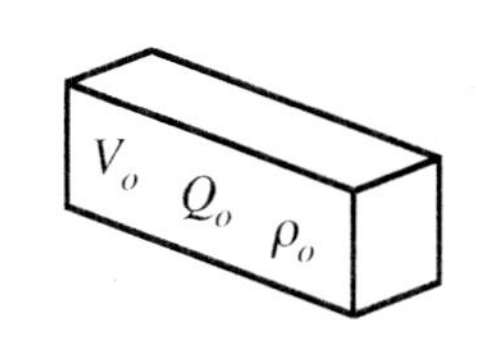

The block of insulating material shown to the right has a volume V_o. An overall charge Q_o is spread evenly throughout the volume of the block so that the block has a uniform charge density ρ_o.

Four charged insulating blocks are shown below. For each block, the volume is given as well as *either* the charge or the charge density of the block.

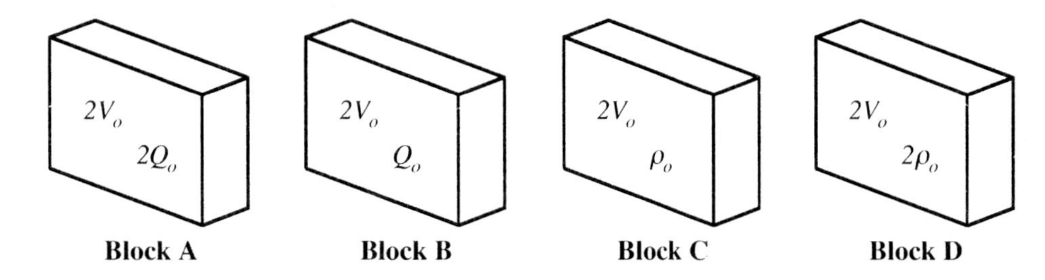

Construct bar charts for (1) the charge and (2) the charge density for the original block, the blocks labeled A – D, and for pieces of the blocks if they were cut in half.

Charge

Original block | Block A | Block B | Block C | Block D | (Block A)/2 | (Block B)/2 | (Block C)/2 | (Block D)/2

Charge Density

Original block | Block A | Block B | Block C | Block D | (Block A)/2 | (Block B)/2 | (Block C)/2 | (Block D)/2

ET3-BCT1: THREE CHARGES IN A LINE—FORCE

Two charged particles, *A* and *B*, are fixed in place. A third charge, *C*, is fixed in place to the right of charge *B* at twice the distance between *A* and *B*. All charges have the same magnitudes.

Construct a bar chart for the net force on charge *B* due to charges *A* and C. Use positive values for net forces directed to the right and negative values for net forces directed to the left. If the force is zero, state that explicitly.

	A	*B*	*C*
Case A	+	+	+
Case B	+	+	–
Case C	+	–	+
Case D	+	–	–
Case E	–	+	+
Case F	–	+	–
Case G	–	–	+
Case H	–	–	–

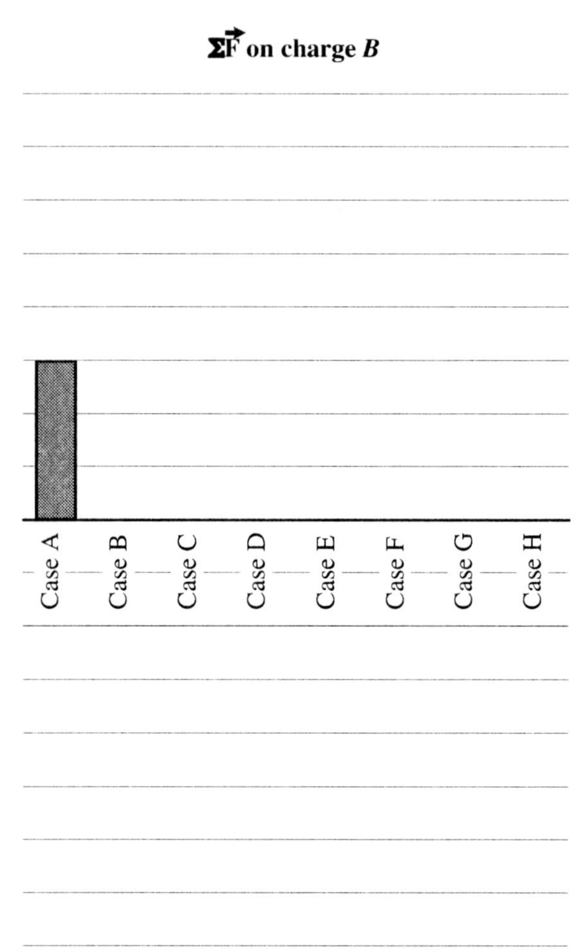

ET5-BCT1: POTENTIAL VS POSITION GRAPH II—ELECTRIC FIELD

Shown below is a graph of potential versus position for a region of space. The labeled points indicate six locations within this region.

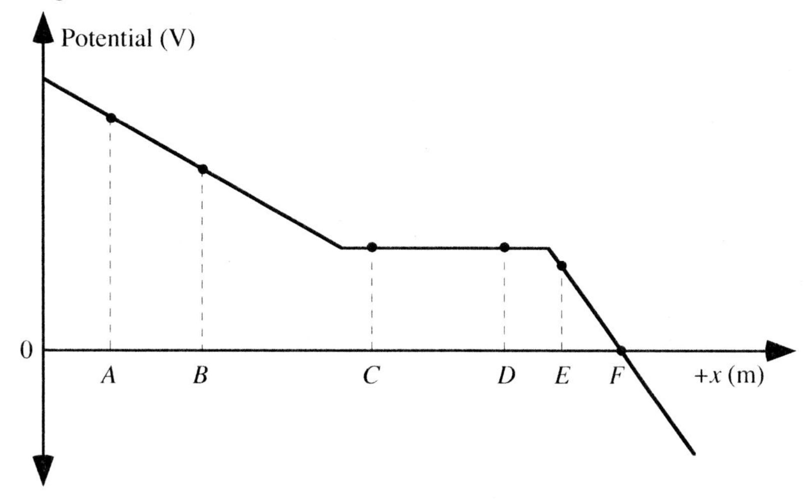

The strength (magnitude) of the x-component of the electric field at point E is indicated in the bar chart shown below.

Complete the sketch of the bar charts for the other points.

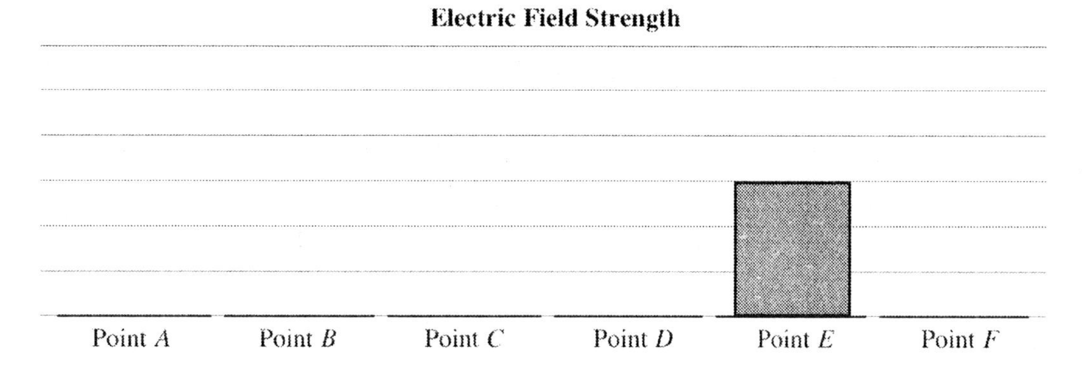

ET5-BCT2: POINT CHARGE—ELECTRIC FIELD

Points *P, R, S,* and *T* lie close to a positive point charge. The concentric circles shown are equally spaced with radii of r, $2r$, $3r$, and $4r$. The magnitude of the electric field at point *P* due to the point charge is shown in the bar chart below.

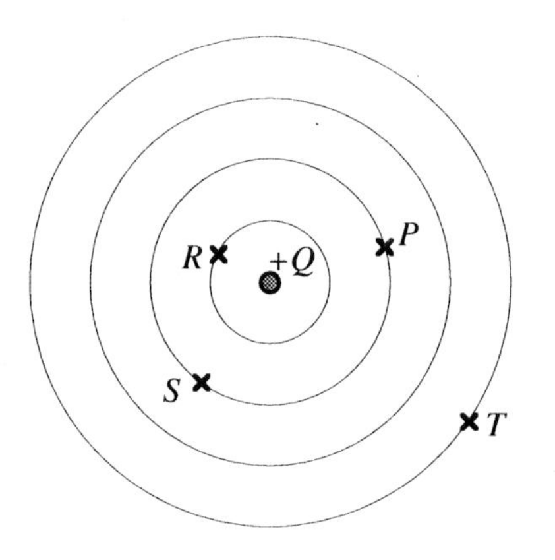

Complete the sketch of the bar charts to indicate the relative magnitude of the electric field at points *R, S,* and *T*.

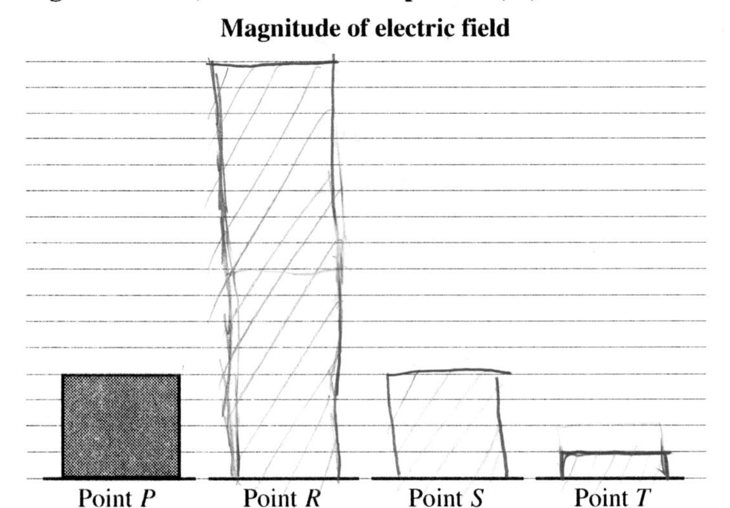

ET5-BCT3: CHARGED CONDUCTING SPHERICAL SHELLS—ELECTRIC FIELD

Three concentric spherical conducting shells are shown in the cross-section to the right.

The inner shell has a net positive charge of 1 μC. The middle shell has a net negative 4 μC charge. The outer shell has a net positive 19 μC charge.

Six points are labeled in the diagram. The distance from the center of the shells to these points is given in the table below, as are the distances to the shell surfaces.

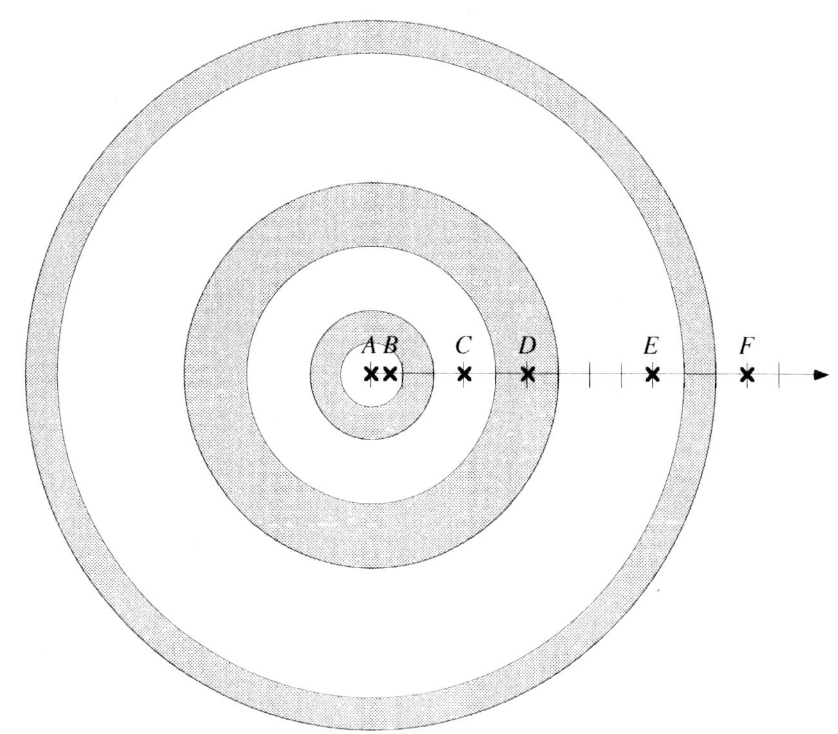

			Inner shell			Middle shell				Outer shell		
	A	*B*	Inside	Outside	*C*	Inside	Outside	*D*	*E*	Inside	Outside	*F*
Distance from center	0	*d*/2	*d*	2*d*	3*d*	4*d*	6*d*	5*d*	9*d*	10*d*	11*d*	12*d*

The electric field at point *C* is shown in the bar chart below. **Indicate the electric field at the other points.** (Positive values indicate a field directed outward; negative values indicate it directed inward).

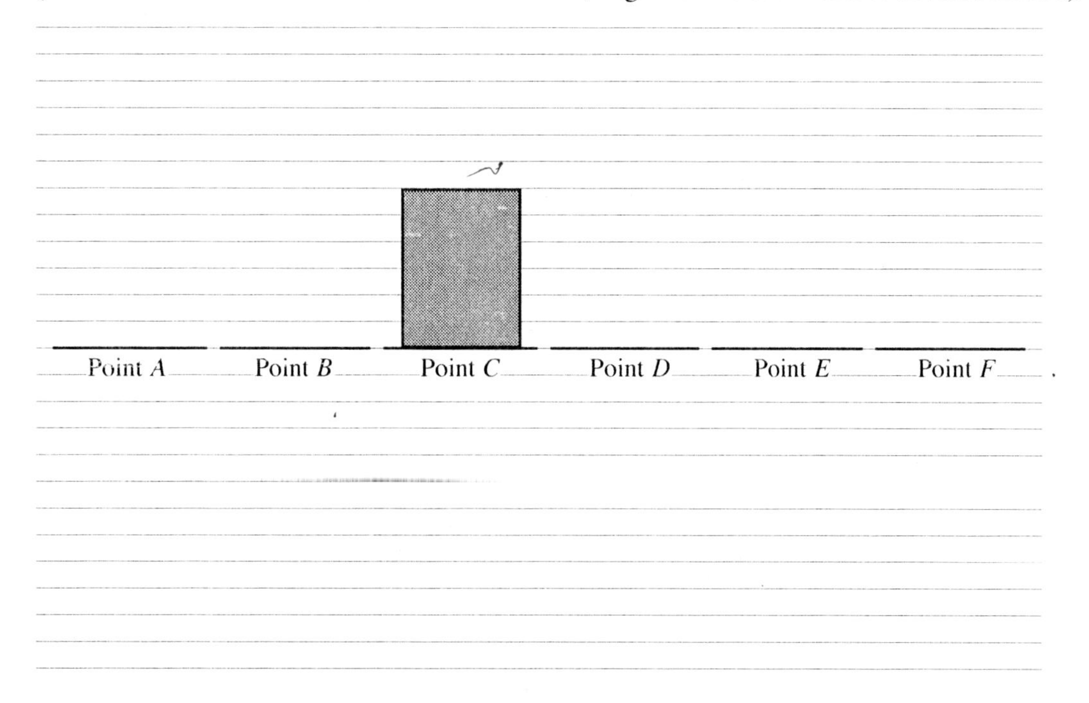

ET7-BCT1: Potential near Two Charges—Electric Field and Potential

Shown below are bar charts for the electric potential and the magnitude of the electric field at the point midway between two equal magnitude charges. The electric potential is zero, and the field has a magnitude of four units.

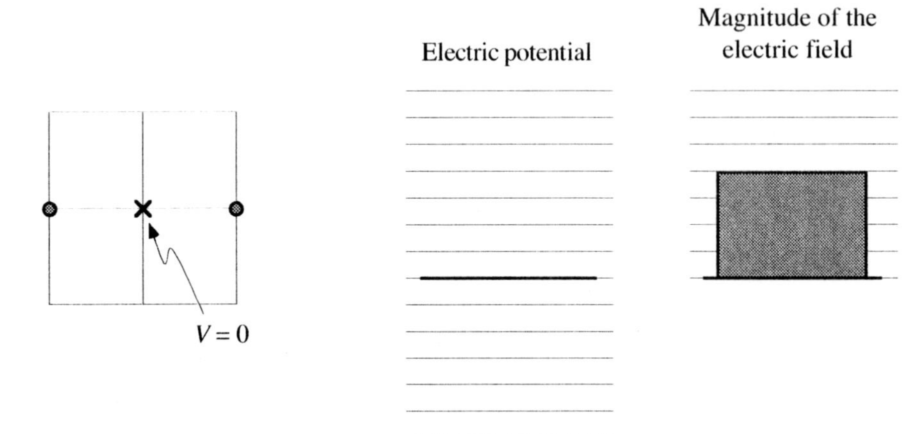

The sign of one of the two charges is changed. **Draw the bar charts for the new electric field magnitude and magnitude of the potential.**

Electric potential

Magnitude of the electric field

ET6-BCT1: SYSTEMS OF POINT CHARGES—WORK TO ASSEMBLE

Shown below are three arrangements of eight positive point charges on identical grids. Each of the point charges has the same charge and the same mass.

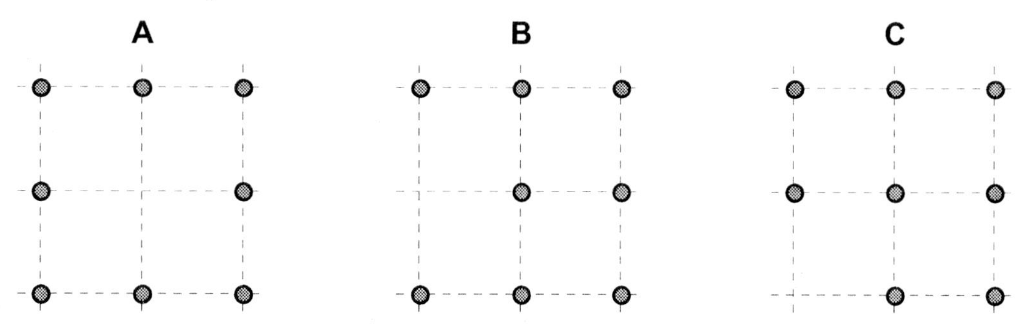

Represent the work to assemble these arrangements with a bar chart sketch. The amount of work is indicated on the bar chart for arrangement A.

A B C

Work required for assembly

WHAT, IF ANYTHING, IS WRONG TASKS (WWT)

ET1-WWT1: BREAKING A CHARGED INSULATING BLOCK—CHARGE DENSITY

A block of insulating material has a positive charge $+Q_o$ distributed uniformly throughout its volume. The block is then broken into two pieces, A and B, as shown.

A student makes the following statement:

"The charge density is calculated by dividing the total charge by the volume. Since the volume is in the denominator, a large volume will give a small charge density. Therefore, the block with the smallest volume, block B, will have the largest charge density."

What, if anything, is wrong with this statement? If something is wrong, explain the error and how to correct it. If the statement is valid, explain why.

ET1-WWT2: BREAKING A CHARGED INSULATING BLOCK—CHARGE DENSITY

A block of insulating material has a positive charge $+Q_o$ distributed uniformly throughout its volume. The block is then broken into two pieces, A and B, as shown.

A student makes the following statement:

"When the block is broken up, there is less overall charge in the pieces than there was in the original block. Charge density is proportional to the overall charge, since $\rho = Q/V$, so more charge means more charge density. This means that the original piece had the greatest charge density, since it had all the charge."

What, if anything, is wrong with this statement? If something is wrong, explain the error and how to correct it. If the statement is valid, explain why.

ET4-WWT1: EQUIPOTENTIAL LINES—DIRECTION OF PROTON'S MOTION

A proton is placed at rest in a uniform electric field and then released. The lines in the diagram represent electric equipotentials.

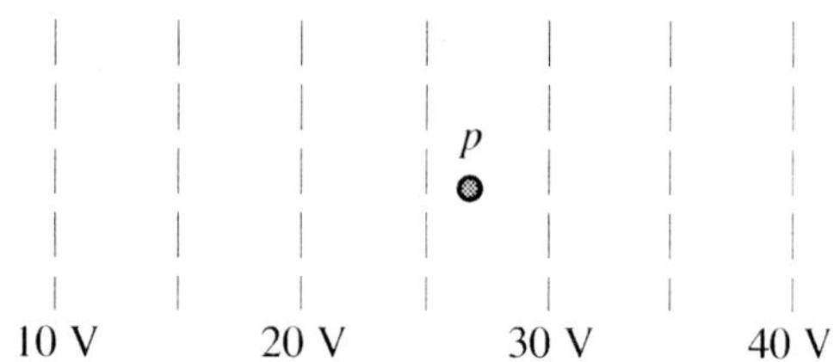

A student makes the following statement:

"A positive particle moves in the direction of lower potential; so, the proton will move to the left."

What, if anything, is wrong with this statement? If something is wrong, explain the error and how to correct it. If the statement is valid, explain why.

ET1-WWT3: INSULATOR AND A GROUNDED CONDUCTOR—INDUCED CHARGE

A charged insulating sphere and a grounded conducting sphere are far apart initially. The charged insulator is then moved to a place near the grounded conductor, as shown in the figure.

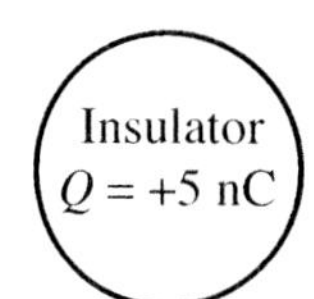

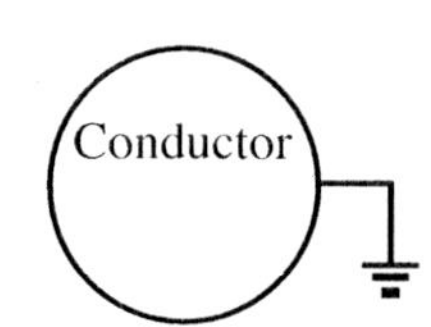

A student makes the following statement:

"When the charged insulator is brought close to the grounded conductor, it will cause the negative charges in the conductor to move to the side closest to the insulator. If the charged insulator is taken away, the conductor will be left with a negative charge evenly distributed over its surface."

What, if anything, is wrong with this statement? If something is wrong, explain the error and how to correct it. If the statement is valid, explain why.

ET1-WWT4: BALLOON STICKING ON A WALL—CHARGE DISTRIBUTION

A student describes how a balloon sticks to a wall and then later falls off.

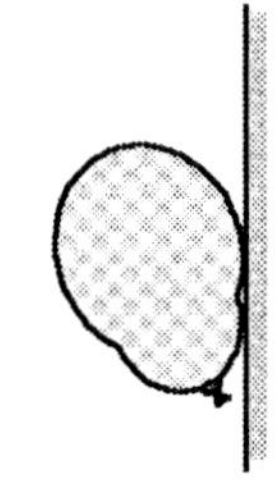

"The wall is a neutral insulator. In the presence of the negatively charged balloon, the positive molecules within the wall are attracted toward the balloon, causing the balloon to stick. The negative molecules in the wall go to the other side, or away from the balloon. The balloon eventually falls because the negative charges of the balloon combine with the positive charges of the wall."

What, if anything, is wrong with this statement? If something is wrong, explain the error and how to correct it. If the statement is valid, explain why.

ET1-WWT5: NEUTRAL METAL SPHERE WITH A CHARGED ROD—CHARGE DISTRIBUTION

A student observes a demonstration involving an interaction between a neutral metallic sphere suspended from a string and a negatively charged insulating rod. The student makes the following statement:

"As the negatively charged rod nears the sphere, it causes that side of the sphere nearest the rod to gain an excess of positive charge while the other side gains an excess of negative charge. So the sphere will be attracted towards the rod. If they touch, the sphere will swing back since they will both become neutral."

What, if anything, is wrong with this statement? If something is wrong, explain the error and how to correct it. If the statement is valid, explain why.

ET5-WWT1: ELECTRIC FORCE ON AN ELECTRON—ELECTRIC FIELD

The graph below shows the electric force acting on an electron at different times.

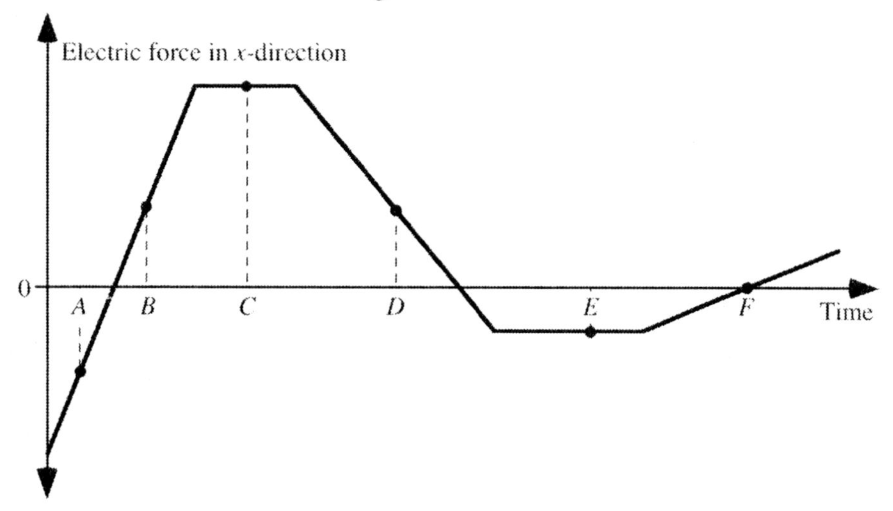

A student makes the following contention about this graph:

"The electric fields at times A and B are equal because the slope of the line is the same. Also, the fields at times C and E are both equal to zero because the slope is zero. The electric field at time D is pointing in the negative direction while the other directions are positive or zero."

What, if anything, is wrong with this statement? If something is wrong, explain the error and how to correct it. If the statement is valid, explain why.

ET5-WWT2: HOLLOW CONDUCTORS—FIELD

A particle of charge $+Q_0$ rests at the center of two hollow concentric conductors of radii R_1 and R_2. Neither conducting shell has a net charge.

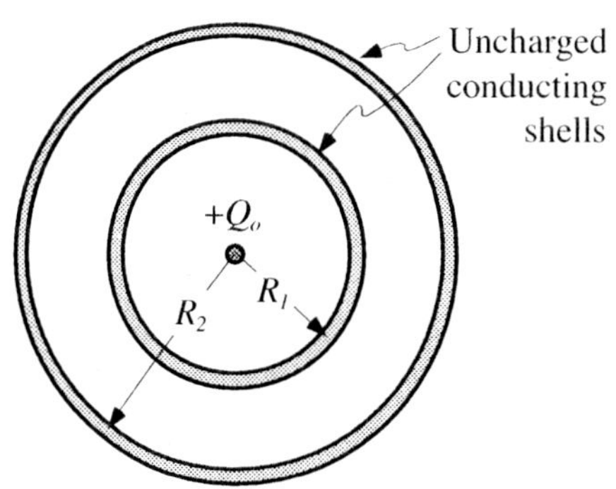

A student makes the following statement:

"The electric field is zero everywhere outside the inner shell because the inner conducting sphere blocks the electric field of Q_0. Thus, the outer sphere has no charge on its surface."

What, if anything, is wrong with this statement? If something is wrong, explain the error and how to correct it. If the statement is valid, explain why.

ET3-WWT1: CHARGES ARRANGED IN A TRIANGLE—FORCE

Three charges are arranged in an isosceles triangle as shown at right. A student trying to find an expression for the net force on the charge Q due to two other charges, each of magnitude q, arrives at the following expression:

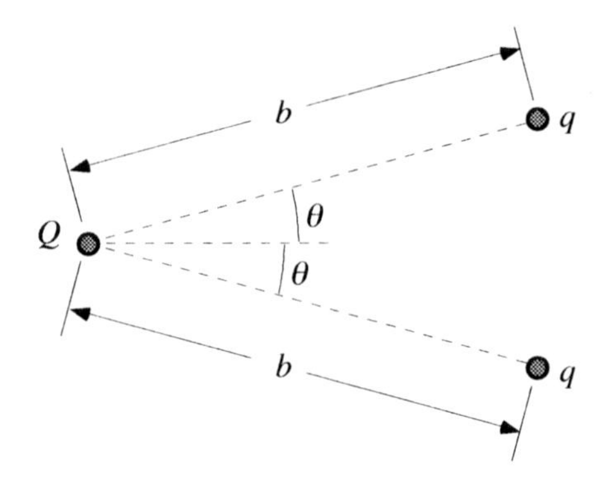

$$F = 2k\frac{Qq}{b^2}$$

What, if anything, is wrong with this expression? If something is wrong, explain the error and how to correct it. If the statement is valid, explain why.

ET3-WWT2: TWO CHARGES—FORCE

Two negatively charged particles are separated by a distance x. The particle on the left has a charge Q which is three times the charge q of the particle on the right.

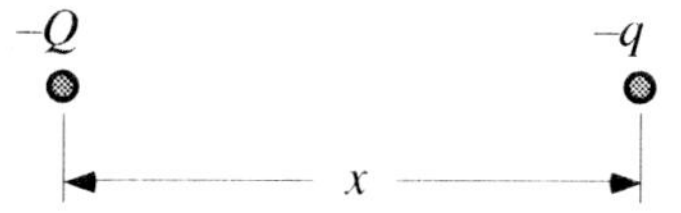

A student makes the following statement:

"Since $F = kQq/x^2$ and Q and q are both negative, the force on Q will be positive. Therefore the force on Q points to the right."

What, if anything, is wrong with this statement? If something is wrong, explain the error and how to correct it. If the statement is valid, explain why.

ET3-WWT3: TWO CHARGED OBJECTS—FORCE

Shown below is a student's diagram for the electric forces acting on two positively charged (q and $4q$) small objects. The objects have the same mass.

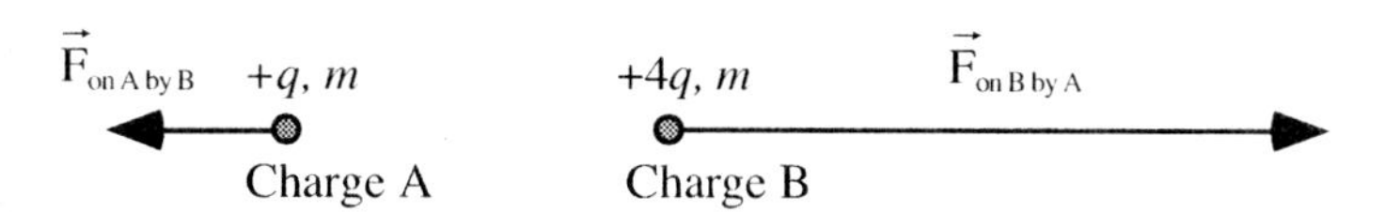

What, if anything, is wrong with this diagram? If something is wrong, explain the error and how to correct it. If the diagram is valid, explain why.

ET3-WWT4: Straight Charged Rod and Two Point Charges—Force

A student considers a situation where a point charge is sitting midway between a +10 nC point charge and a rod with a uniform charge distribution.

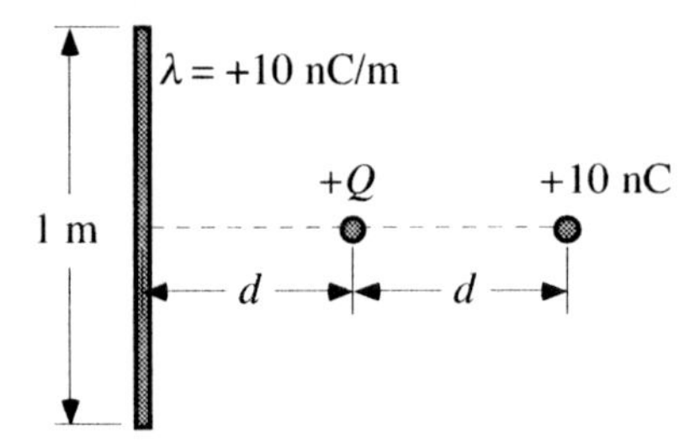

The student makes the following statement:

"The net force on $+Q$ will be zero since the point charge and the rod have the same charge so they will exert forces on $+Q$ that are equal in strength but oppositely directed."

What, if anything, is wrong with this statement? If something is wrong, explain the error and how to correct it. If the statement is valid, explain why.

ET3-WWT5: Sphere and a Point Charge—Force

A positive point charge is placed a distance d away from a neutral metal sphere.

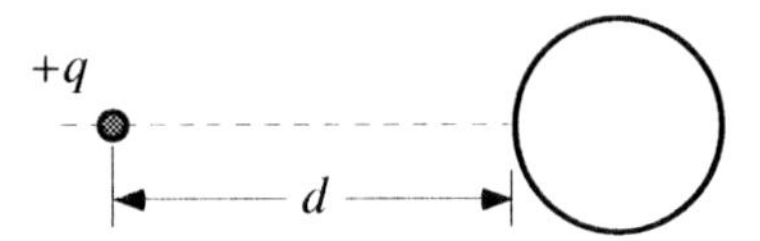

A student makes the following statement:

"The electric force is zero. Coulomb's law states that the electric force between two objects is proportional to the product of the charges. Since the charge of the sphere is zero, and zero times anything gives zero, the force between the point charge and the sphere is zero."

What, if anything, is wrong with this statement? If something is wrong, explain the error and how to correct it. If the statement is valid, explain why.

ET3-WWT6: UNIFORM ELECTRIC FIELD—ELECTRIC FORCE

The figure below displays a student drawing of an electric *force* vs. *time* graph of a negatively charged particle released from rest in a uniform electric field.

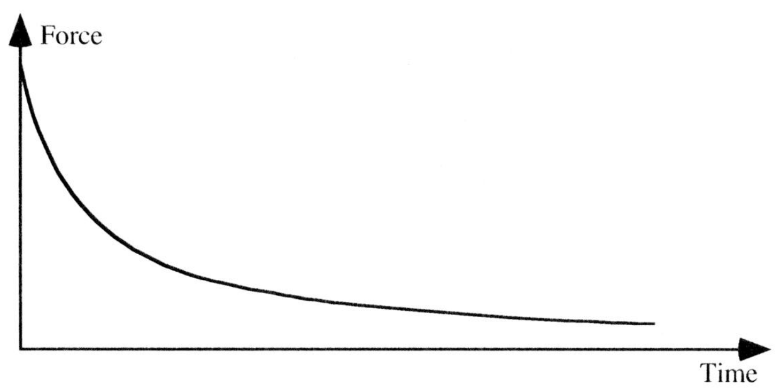

What, if anything, is wrong with the graph above? If something is wrong, explain the error and how to correct it. If the graph is legitimate, explain why.

ET4-WWT2: ELECTRON IN A UNIFORM ELECTRIC FIELD—VELOCITY

An electron is placed with an initial velocity of 5 m/s in a uniform electric field as shown below.

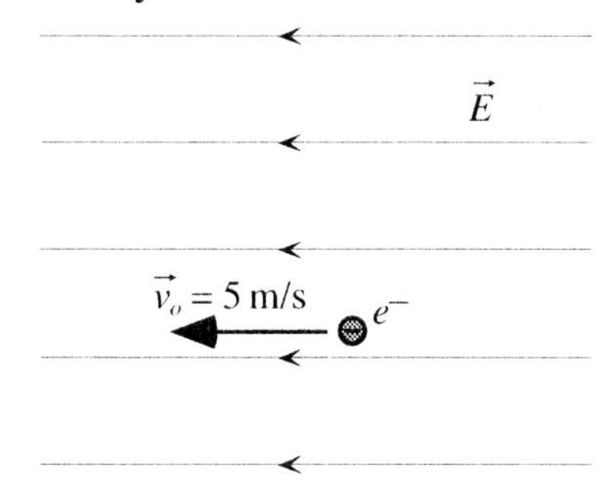

A student makes the following statement:

"The electron will continue to move in the same direction at a constant velocity because it is moving in the same direction as the electric force on it; since the electric field is constant, the force on the electron is constant."

What, if anything, is wrong with this statement? If something is wrong, explain the error and how to correct it. If the statement is valid, explain why.

ET5-WWT3: POTENTIAL NEAR TWO CHARGES—ELECTRIC FIELD

The electric potential at the midpoint between two equal magnitude electric charges is zero.

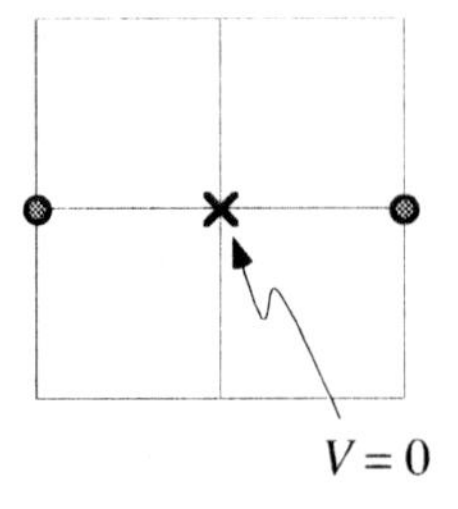

A student considering this situation says:

"The electric field at the midpoint between the two charges will be zero. Since the two charges are opposite in sign, the fields will be equal but opposite, and add to zero."

What, if anything, is wrong with this statement? If something is wrong, explain the error and how to correct it. If the statement is valid, explain why.

ET5-WWT4: THREE CHARGES IN A LINE—ELECTRIC FIELD

Three point charges with magnitudes and signs given in the figure are placed in a row.

+8 mC +6 mC –3 mC

q_1 q_2 q_3

x x

A student makes the following statement:

"If you double the magnitude of the middle charge, the force on that charge doubles due to the other charges by Coulomb's law. Since $F = qE$, if we double the force, the field doubles also. So the field that acts on the middle charge doubles."

What, if anything, is wrong with this statement? If something is wrong, explain the error and how to correct it. If the statement is valid, explain why.

ET5-WWT5: POTENTIAL VS POSITION GRAPH II—ELECTRIC FIELD

A graph is shown of potential versus position for a region of space. The labeled point represents a location within this region.

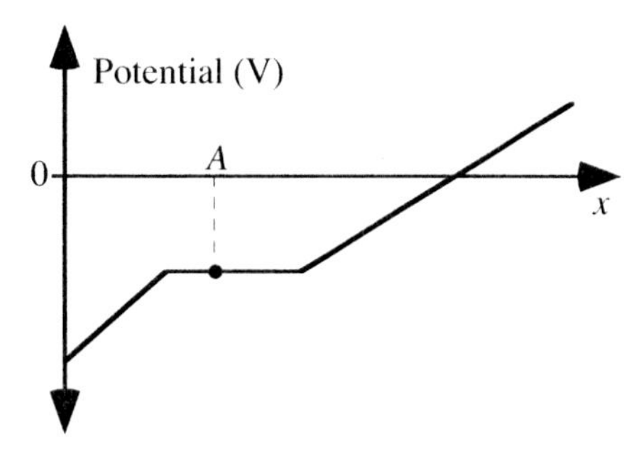

A student makes the following contention about this graph:

"At point A, the x-component of the electric field will be zero since the line is horizontal around that point."

What, if anything, is wrong with this statement? If something is wrong, explain the error and how to correct it. If the statement is valid, explain why.

ET5-WWT6: FIELD OUTSIDE A SPHERE WITH A CAVITY—ELECTRIC FIELD

Shown is a cross-section of a conducting sphere that has a spherical cavity in it. Fixed in position in the cavity is a positive charge $+Q$. Points P and S are both 1 cm away from the surface of the sphere.

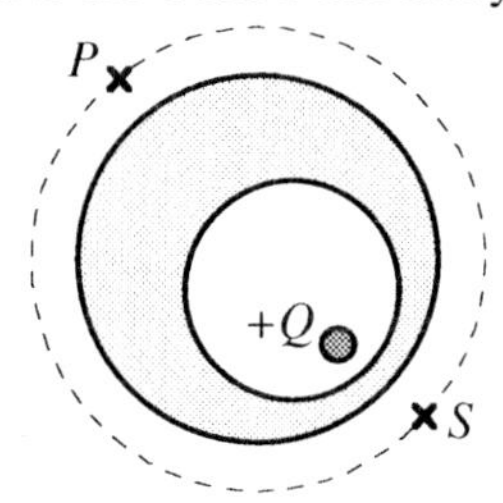

A student considering this situation makes the following statement:

"The electric field at point S will be larger than the field at point P, since there will be more charge on the surface near the interior charge compared to farther away."

What, if anything, is wrong with this statement? If something is wrong, explain the error and how to correct it. If the statement is valid, explain why.

ET6-WWT1: Moving Charged Particle in an Electric Field—Potential Energy

At the instant shown, a positively charged particle is moving at 5 m/s in the direction of a uniform electric field.

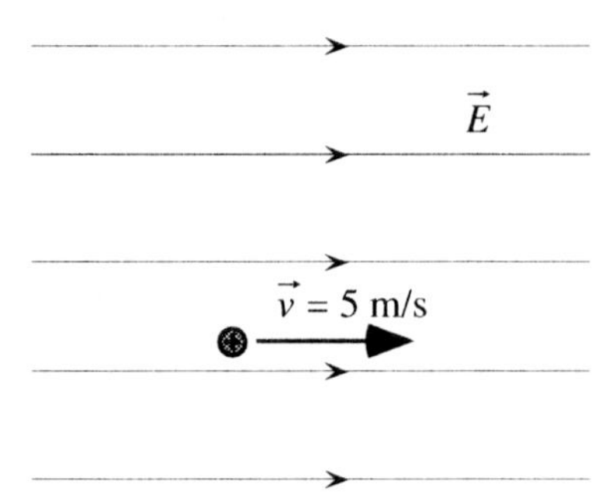

A student makes the following statement:

"As the particle continues to move in the direction of the electric field, it will gain potential energy since it is moving in the direction of increasing electric potential."

What, if anything, is wrong with the above statement? If something is wrong, explain the error and how to correct it. If the statement is valid, explain.

ET8-WWT1: Uniformly Charged Insulating Sphere—Electric Potential

Points A and B are inside of a uniformly-charged, solid, insulating sphere with a total charge Q.

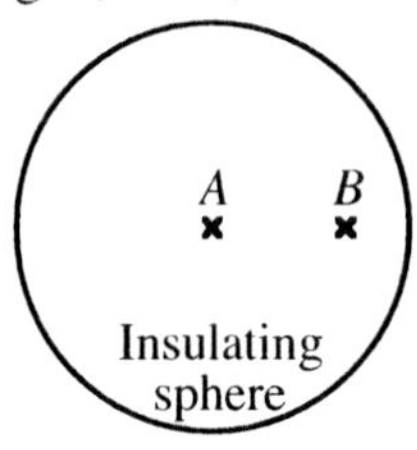

A student makes the following statement:

"The electric potential at point A is equal to the electric potential at point B because the electric potential is always the same inside a sphere."

What, if anything, is wrong with this statement? If something is wrong, explain the error and how to correct it. If the statement is valid, explain why.

ET8-WWT2: Two Large Charged Parallel Sheets—Potential Difference

The diagram shows two very large parallel charged insulating sheets. (Only a small portion near the center of the sheets is shown; the distance between the sheets is very small compared to the dimensions of the sheets.) The sheets are uniformly charged with charge densities and separation as shown.

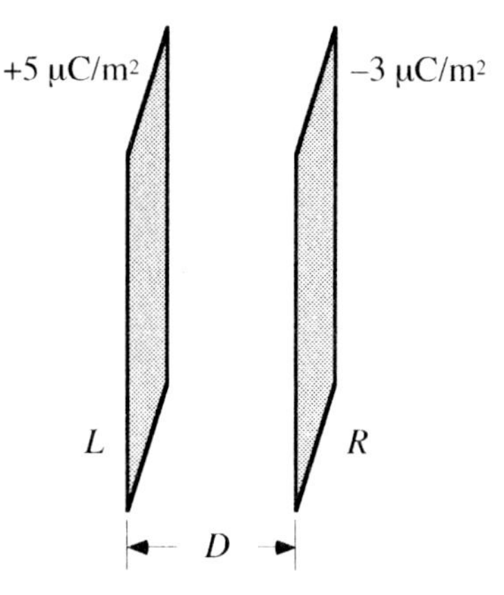

A student makes the following statement:

"The electric field due to a large sheet of charge is independent of the distance from the sheet. The electric field between the two sheets will be the superposition of the electric fields from the individual sheets. So the field between the sheets is constant. The electric potential between the sheets is therefore constant, and there is no electric potential difference between the sheets."

What, if anything, is wrong with this statement? If something is wrong, explain the error and how to correct it. If the statement is valid, explain why.

ET9-WWT1: Uniform Electric Field—Electric Flux

The figure below displays a Gaussian cube with sides of 10 cm inside a uniform electric field of 400 N/C in the direction shown.

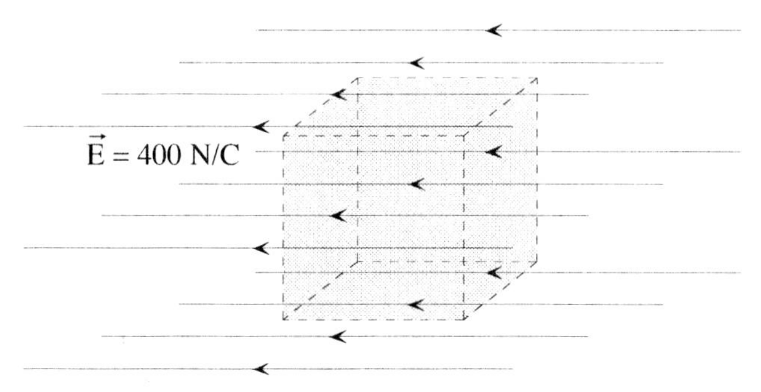

Given below is a student's calculation for the net electric flux through the cube.

$$\Phi_{Net} = 24\ \text{Nm}^2/\text{C} = (6\ \text{sides})(E)(\text{Area of one side}) = (6)(400\ \text{N/C})(10.0\ \text{cm})^2$$

What, if anything, is wrong with this calculation? If something is wrong, explain the error and how to correct it. If the calculation is valid, explain why.

TROUBLESHOOTING TASKS (TT)

ET3-TT1: CHARGES ARRANGED IN A TRIANGLE—FORCE

Three charges are arranged in an isosceles triangle as shown. A student trying to find the magnitude of the net force on charge Q due to the other two charges arrives at the following expression:

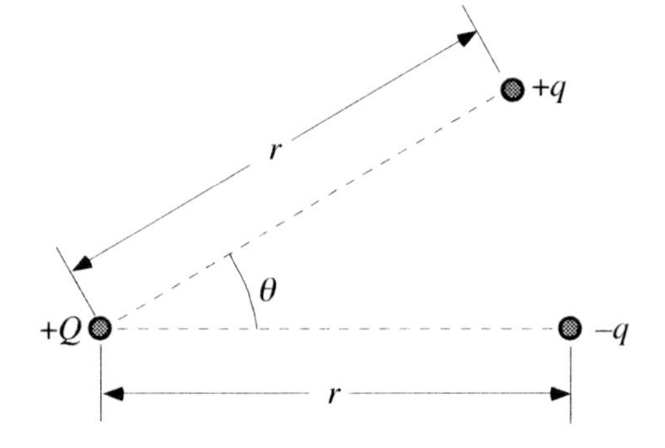

$$|F| = -k\frac{Qq}{r^2} + k\frac{Qq}{r^2}\cos\theta$$

There is something wrong with the student's expression. **Identify any problems and explain how to correct them.**

ET3-TT2: TWO CHARGED OBJECTS—FORCE

Shown below is a student's drawing of the electric forces acting on object A (with charge $+q$ and mass m) and object B (with charge $+4q$ and mass $4m$).

$\vec{F}_{\text{on A by B}}$ $+q$, m — Object A

$+4q$, $4m$ — Object B $\vec{F}_{\text{on B by A}}$

There is something wrong with this diagram. **Please explain how to correct it.**

ET3-TT3: STRAIGHT CHARGED ROD AND TWO POINT CHARGES—FORCE

A point charge, $+Q$, is sitting midway between a +10 nC point charge and a rod with a uniform charge distribution.

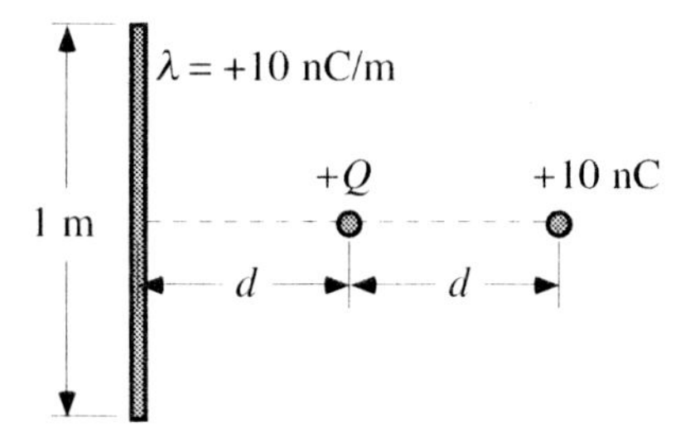

The student makes the following statement:

"The net force on +Q will be zero. Since the two charges have the same magnitude, they will exert forces on +Q that are equal in strength, but oppositely directed."

There is something wrong with the student's contention. **Identify any problem(s) and explain how to correct it/them.**

ET3-TT4: SPHERE AND A POINT CHARGE—FORCE

A positive point charge is placed a distance d away from a neutral solid metal sphere.

A student makes the following statement about the electric force between the neutral metal sphere and the point charge:

"There is an attraction between the point charge and the sphere. Since the sphere is a conductor, the external positive point charge pulls electrons in the sphere toward it. This leaves positive charges on the other side of the sphere, since the sphere is still neutral. The force between the point charge and the sphere is just the attraction between the negative charges on the left end of the sphere and the point charge."

There is at least one problem with this student's contention. **Identify any problem(s) and explain how to correct it/them.**

ET4-TT1: ELECTRON MOVING INTO A UNIFORM ELECTRIC FIELD—ACCELERATION

An electron is shot into a uniform electric field that is directed upward.

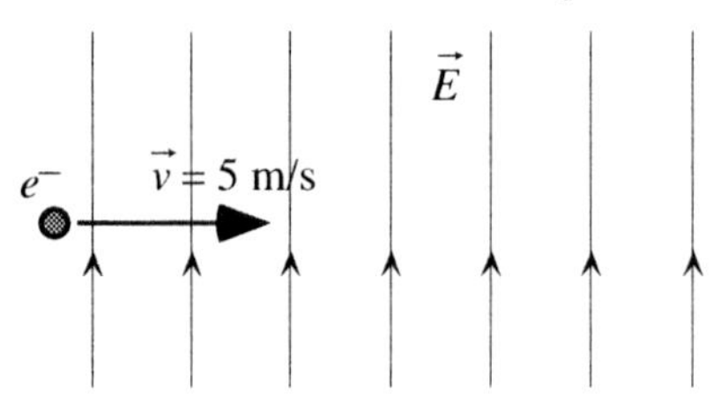

There is something wrong with the statement below about this situation.

"Since an electron is negative, it will accelerate downward, or in the opposite direction of the electric field. This acceleration is perpendicular to the motion, so it will not increase the magnitude of the velocity, just change its direction."

What is wrong and how could it be corrected?

ET5-TT1: POTENTIAL VS POSITION GRAPH II—ELECTRIC FIELD

A graph of electric potential versus position is shown.

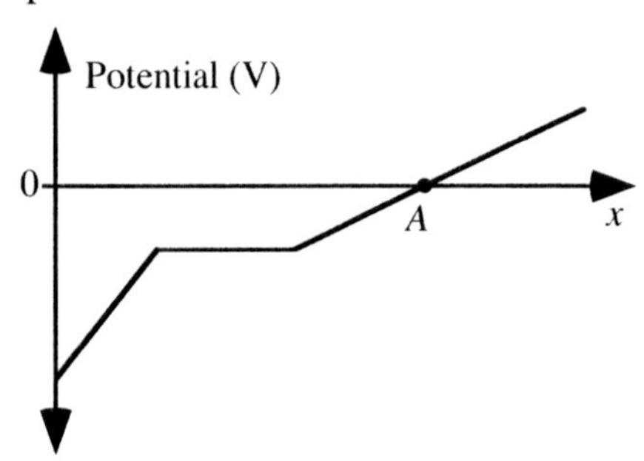

A student considering this graph makes the following contention:

"The electric field at point A is zero since that is the point at which the potential reverses from negative to positive."

There is a problem with this student's contention. **Identify any problem(s) and explain how to correct it/them.**

ET5-TT2: POTENTIAL NEAR TWO CHARGES—ELECTRIC FIELD

Two equal magnitude electric charges are separated by a distance d. The electric potential at the midpoint between these two charges is zero.

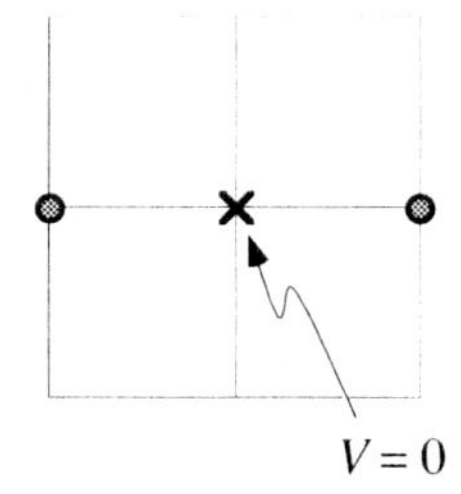

A student considering this situation says:

"The electric field at the midpoint between the two charges will be zero also, since the two charges are opposite in sign, so the fields will be equal but opposite, and add to zero."

There is something wrong with the student's statement. **Identify any problem(s) and explain how to correct it/them.**

ET5-TT3: THREE CHARGES IN A LINE—ELECTRIC FIELD

Three charges are placed in a row. Adjacent charges are 2 cm apart, with magnitudes and signs given in the figure.

+8 mC　　+6 mC　　–3 mC

q_1　　q_2　　q_3

2 cm　　2 cm

A student makes the following calculation for the magnitude of the electric field acting on the middle charge:

$$E = k\frac{q_1}{r_1^2} + k\frac{q_2}{r_2^2} + k\frac{q_3}{r_3^2} = 9x10^9\ Nm^2/C^2\left[\frac{8mC}{(2cm)^2} + \frac{6mC}{(2cm)^2} - \frac{3mC}{(2cm)^2}\right]$$

There is at least one problem with the student's calculation. **Identify the problem(s) and explain how to correct it/them.**

ET7-TT1: Two Connected Charged Spheres—Potential and Charge

Two charged, conducting spheres of different sizes are connected by a wire.

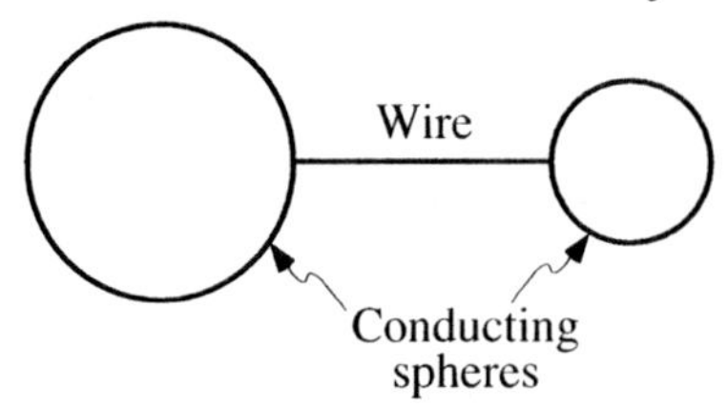

A student considering the situation says:

"Since the two spheres are connected together, they will have the same electric potential and the charges on the surfaces of the spheres will be equal as well."

There is something wrong with the statement above. **Identify any problem(s) and explain how to correct it/them.**

ET8-TT1: Two Large Charged Parallel Sheets—Potential Difference

The diagram at right shows two very large parallel insulating sheets. (Only a small portion near the center of the sheets is shown; the distance between the sheets is very small compared to the dimensions of the sheets.) The sheets are uniformly charged with charge densities and separation as shown.

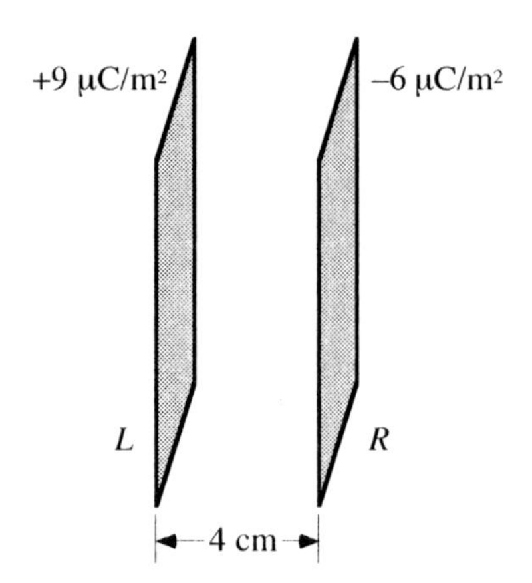

A student makes the following comment about what would happen to the electric potential difference between the sheets if the sheets were moved closer together:

"The electric field between the sheets would increase because the field lines would become more concentrated. The electric potential difference between the sheets depends on the strength of the electric field between the sheets, so this would increase."

There is at least one problem with this student's contention. **Identify any problem(s) and explain how to correct it/them.**

ET9-TT1: Conducting Shell—Electric Flux

The diagram below shows a hollow neutral conductor of inner radius B and outer radius C surrounding a uniformly charged insulating sphere of radius A with a total charge $+Q$.

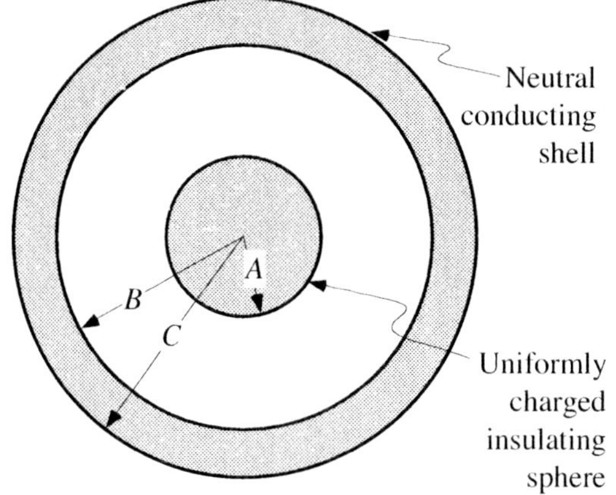

There is something wrong with the description below.

"The electric flux through a Gaussian sphere of radius r, where $B < r < C$ will be the same as the electric flux through a Gaussian sphere of radius R where $R > C$ since the hollow conductor is neutral."

Identify any problem(s) and explain how to correct it/them.

ET10-TT1: Charged Rod near a Suspended Bar Magnet—Rotation Direction

A bar magnet is suspended from a string. With the magnet held in place, a charged rod is brought close to the point of suspension of the magnet as shown. The suspended bar magnet is then released so that it is free to rotate.

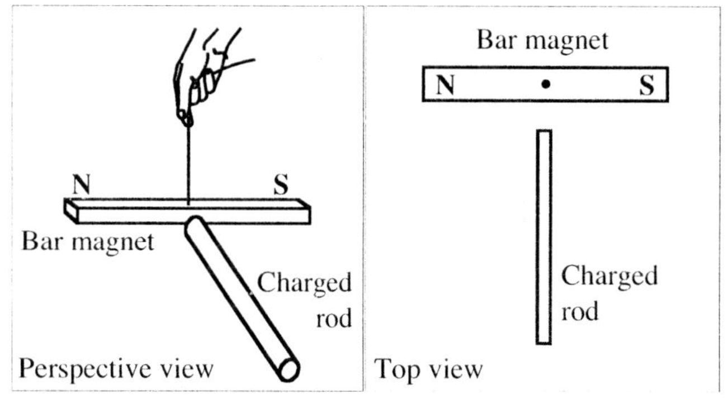

A student makes the following claim about what will happen when the magnet is free to rotate:

"The north pole of the magnet will be repelled by the positively charged rod. The south pole will be attracted. So the magnet will rotate clockwise when viewed from above."

There is something wrong with the student's contention. **Identify any problem(s) and explain how to correct it/them.**

PREDICT AND EXPLAIN TASKS (PET)

ET1-PET1: TWO INSULATING RODS—CHARGE DENSITY

Two charged insulating cylindrical rods are shown. Rod A has a length H and a radius R and rod B has a length $2H$ and a radius $2R$. Both rods have the same total charge. Rod A has a charge density ρ_A.

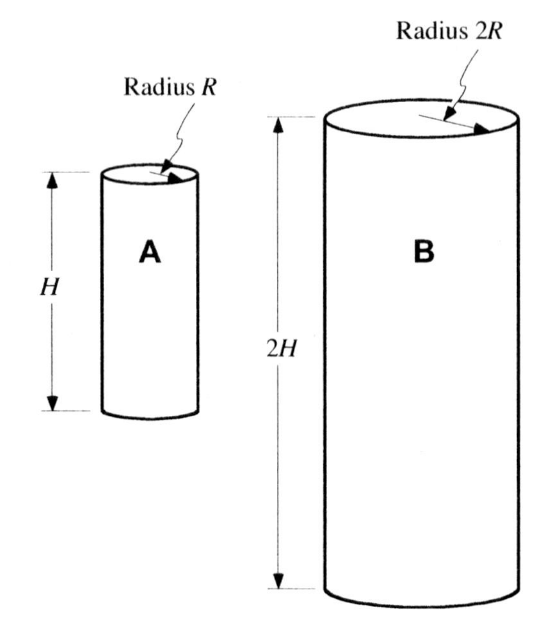

What is the charge density of Rod B in terms of ρ_A? Show and explain your work.

What is the linear charge density (*i.e.*, the charge per unit length) of Rod A in terms of the given variables (ρ_A, R, and H)? Show and explain your work.

What is the linear charge density (*i.e.*, the charge per unit length) of Rod B in terms of the given variables (ρ_A, R, and H)? Show and explain your work.

eT1-PET2: Electroscope—Charge

A student first holds a positively charged rod near the top plate of an electroscope without touching it. The electroscope foil deflects. The electroscope was initially uncharged.

Predict whether the electroscope is now *positively charged, negatively charged,* or *neutral.* Explain.

She then touches the electroscope plate while keeping the positively charged rod near the plate. The electroscope foil falls back to its undeflected position.

Predict whether the electroscope is *positively charged, negatively charged,* or *neutral.* Explain.

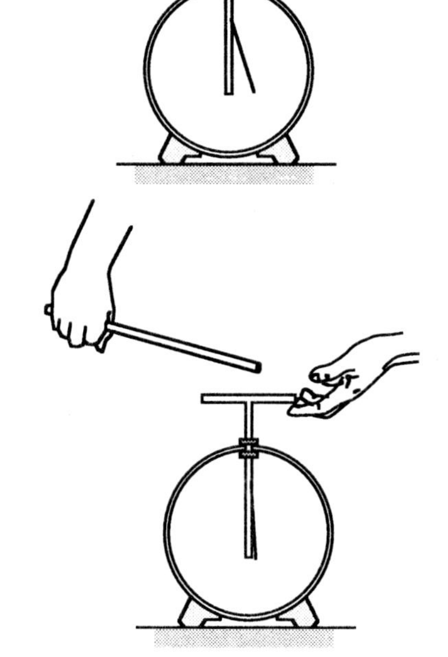

While holding the positively charged rod stationary, she removes her hand which is touching the electroscope. Finally, she removes the charged rod.

Will the electroscope foil be *deflected* or *undeflected*? Explain.

Predict whether the electroscope is *positively charged, negatively charged,* or *neutral.* Explain.

ET3-PET1: Two Charged Objects—Force

Two small, charged objects were initially placed at rest at the positions shown and then released.

1. Predict and explain the motion of each object if they have the same mass (*i.e.*, indicate the relative change in position, their relative velocities, and acceleration).

2. Predict and explain the motion of each object if object *A* has mass m, and object *B* has mass 4m.

ET3-PET2: Conducting Cube Between Point Charges—Force

Two equal and opposite charges are separated by a distance d. The magnitude of the electric force on the positive charge by the negative charge is measured to be F_o.

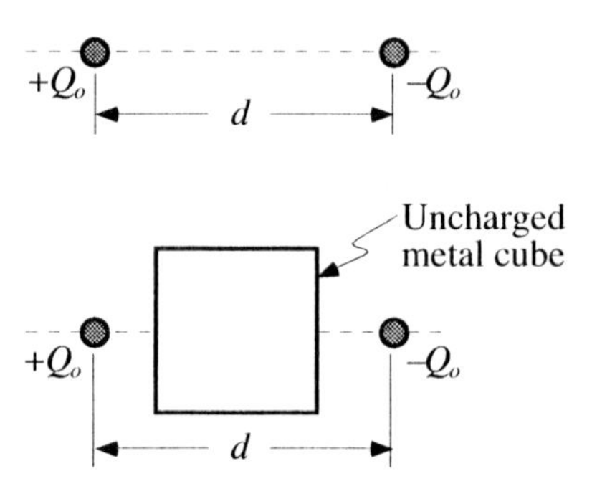

An uncharged metal cube is placed between the two point charges as shown in the lower figure, with the charges fixed so that the distance between them remains the same.

Predict whether the net electric force on the positive charge due to the metal cube and the negative charge is *greater than*, *less than*, or *equal to* F_o. Explain.

ET4-PET1: Straight Charged Rod and Two Point Charges—Acceleration

A point charge $+Q$ is placed midway between a + 10 nC point charge and a rod with a uniform charge distribution.

$\lambda = +10$ nC/m

$+Q$ +10 nC

1 m

d d

Predict which way $+Q$ will accelerate initially if released. Explain.

ET4-PET2: Electric Potential vs Position Graph II—Motion of Charged Particles

Shown below is a graph of potential versus position for a region of space. The labeled points indicate six locations within this region.

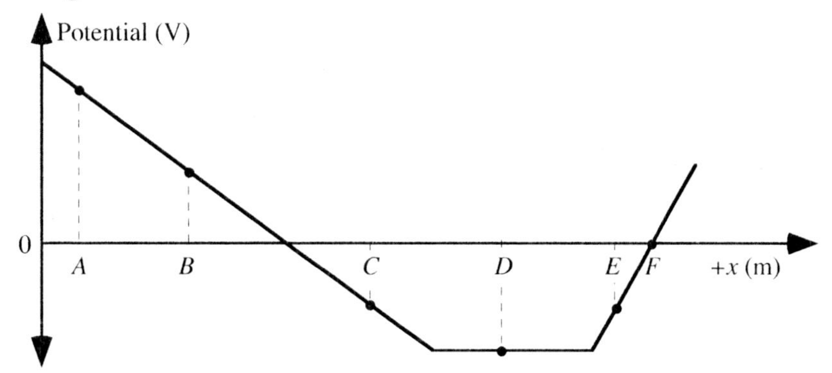

Charged particles are released from rest at various locations in the region one at a time. **Describe how each particle will move initially (if they will) and why.**

Particle type	Released at rest from:	Will it move?	Which direction?	Explain.
positive	A			
positive	C			
positive	E			
positive	F			
negative	B			
negative	D			
negative	E			
negative	F			

ET8-PET1: PARALLEL PLATE CAPACITOR—POTENTIAL

The capacitor in the figure below has a cross-sectional area A_o and a distance between plates of 10 mm. Assume the plates are large enough that the field is uniform in the region of interest. A 30-volt battery is connected across the plates of the capacitor. There is nothing between the plates of the capacitor. The dotted line *AB* is 1 mm from the top plate and the dotted line *CD* is 8 mm from the top plate. The dotted lines are 10 mm long. Assume that $V = 0$ at the lower plate.

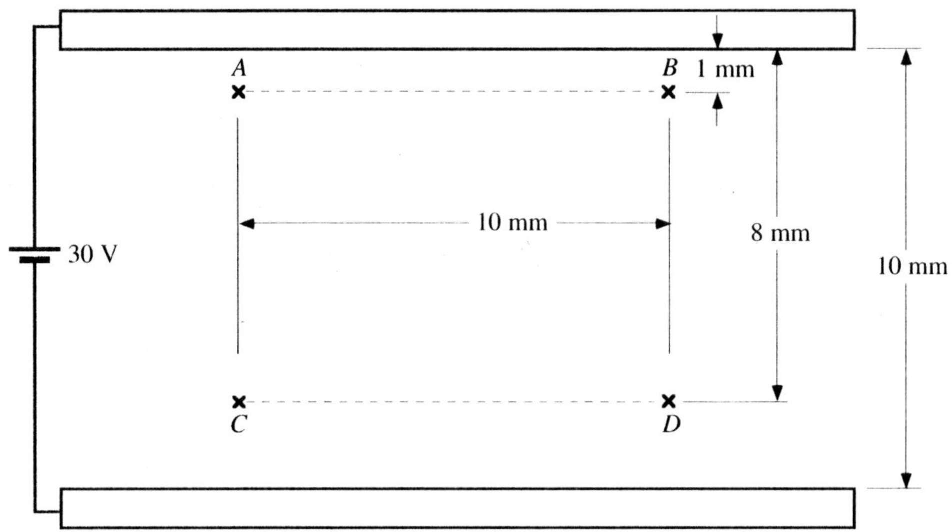

What is the potential at:

***B*________**

***C*________**

Explain.

If the voltage of the battery is doubled to 60 volts, what is the potential at:

***A*________**

***D*________**

Explain.

WORKING BACKWARDS TASKS (WBT)

ET2-WBT1: THREE CHARGES—PHYSICAL SITUATION

Three positive charges with magnitudes 9 mC, 6 mC, and 3 mC are fixed in place. A student makes the following calculation for the magnitude of the electric field acting on the 9 mC charge:

$$|E| = (9 \times 10^9 N \cdot m^2/C^2)\left[\frac{6mC}{(2cm)^2} - \frac{3mC}{(2cm)^2}\right]$$

Draw the arrangement of these three charges and clearly label the point that satisfies this calculation. Explain how your drawing is consistent with the calculation.

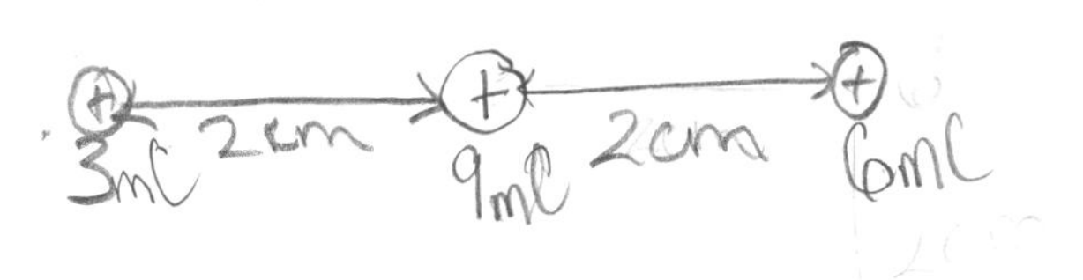

ET2-WBT2: CHARGE ARRANGEMENT—PHYSICAL SITUATION

A student completes a physics problem involving two identical charges (q) and a third charge (Q). She obtains the following equation for the magnitude of the force on Q:

$$F = 2k\frac{Qq}{a^2}\cos\theta$$

Draw a physical situation that would result in this equation. Be sure to label the charges. Explain how your drawing is consistent with the equation.

ET2-WBT3: Electric Field Graphs—Physical Situation

Shown below is a graph of the electric field versus position for a situation involving charged insulators and/or conductors.

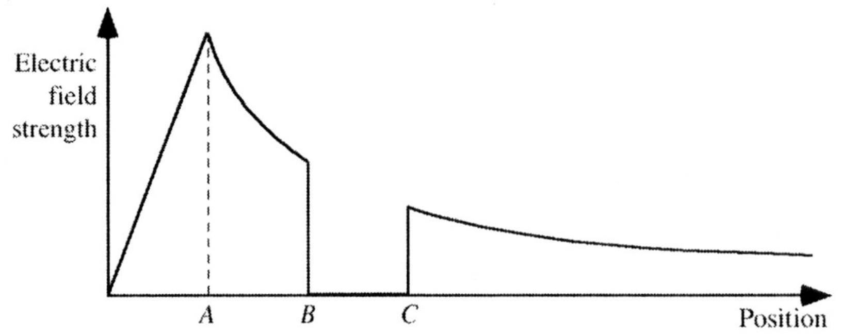

Draw the physical situation that this graph represents and label each object clearly.

ET2-WBT4: Electric Field Graphs—Physical Situation

Shown below is a graph of the electric field versus position for a situation involving charged insulators and/or conductors.

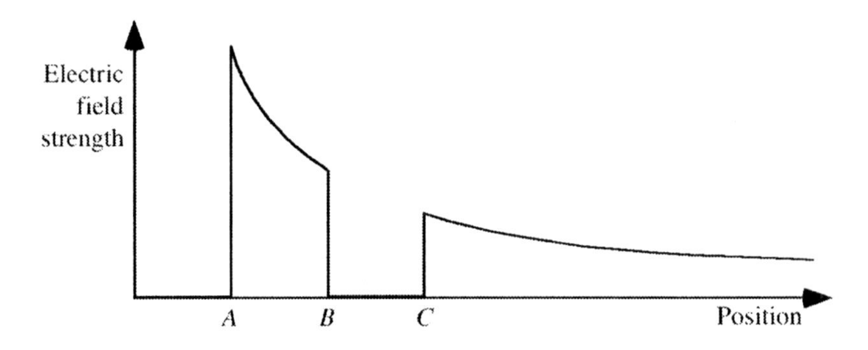

Draw the physical situation that this graph represents and label each object clearly.

ET2-WBT5: ELECTRIC POTENTIAL DIFFERENCE—PHYSICAL SITUATION

A student trying to calculate an electric potential difference makes the following calculation:

$$\Delta V = -\int \vec{E} \bullet d\vec{l} = -Ed = -(E_L + E_R)d = -(\frac{\sigma_L}{2\varepsilon_o} + \frac{\sigma_R}{2\varepsilon_o})d$$

$$\Delta V = -\left[\frac{3\mu C/m^2}{2\varepsilon_o} + \frac{6\mu C/m^2}{2\varepsilon_o}\right](0.08m)$$

Use the calculation above to reconstruct the physical situation. Be sure to label ΔV as part of your diagram.

ET2-WBT6: ELECTRIC POTENTIAL *X* AND *Y* GRAPHS—ELECTRIC FIELD

Shown below are graphs of the electric potential in a region of space with respect to position in the x- and y-directions.

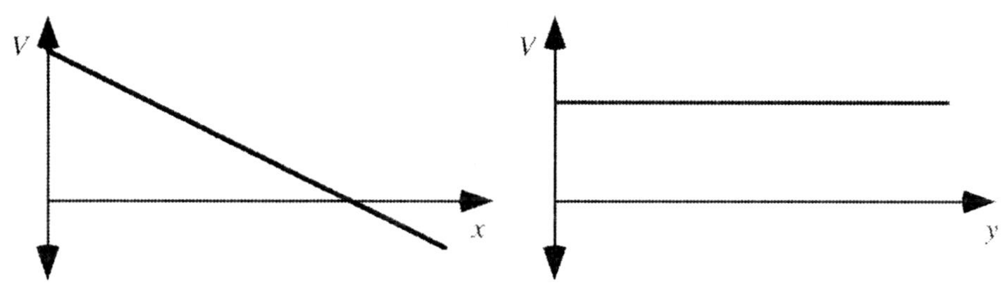

Describe the electric field that is consistent with these graphs.

ET2-WBT7: Charged Rod with Electric Field Components—Length and Location

A student working on a physics problem determines that the x- and y-components of the electric field at a point P near a charged insulating rod can be calculated from the following integrals:

$$E_x = k\lambda \int_0^L \left(\frac{dy}{x^2 + y^2}\right)\left(\frac{x}{\sqrt{x^2 + y^2}}\right)$$

$$E_y = -k\lambda \int_0^L \left(\frac{dy}{x^2 + y^2}\right)\left(\frac{y}{\sqrt{x^2 + y^2}}\right)$$

What is the length of the rod?

Is the rod parallel to the x-axis, to the y-axis, or to neither?

How far from point P is the rod, and where is P in relation to the rod?

Draw the physical situation showing the rod and the point P, and label important features.

ET2-WBT8: POTENTIAL NEAR TWO CHARGES—PHYSICAL SITUATION

The following equations for electric potential and field were determined for a point near two charges.

$$V = k\left[\frac{q}{d} - \frac{q}{d}\right] = 0 \qquad\qquad |E| = k\left[\frac{q}{d^2} + \frac{q}{d^2}\right] = 2k\frac{q}{d^2}$$

Construct and label a configuration of charges including the point that is consistent with these equations.

ET2-WBT9: CHARGED INSULATING SHEETS—ELECTRIC FIELD

Two very large parallel, insulating sheets are separated by 8 cm. (Only a small portion near the center of the sheets is shown.) Sheet *A* has a charge density σ_A. A coordinate system is chosen such that the sheets lie in the *y*-*z* plane, with sheet *A* at $x = 0$. A graph of the electric field as a function of *x* is shown at right below.

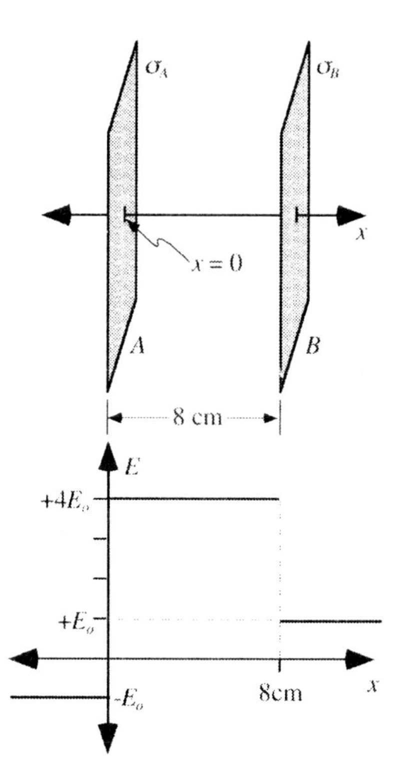

Find the charge density σ_B (sign and magnitude) of sheet *B* in terms of the charge density σ_A of sheet *A*.

ET2-WBT10: FORCES ON THREE CHARGES ALONG A LINE—CHARGE LOCATION

Three charges are fixed in place along a line. All three charges have the same magnitude, but they may have different signs. Shown below are diagrams showing the forces exerted on each charge by the other two charges.

In each case, the sign of one of the charges is shown, as well as its position along a dashed line. **Indicate the signs of the other two charges and their approximate positions on the dashed line.**

Case 1

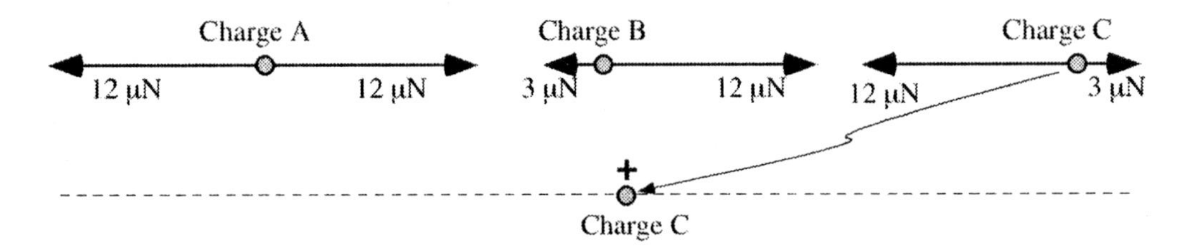

Case 2

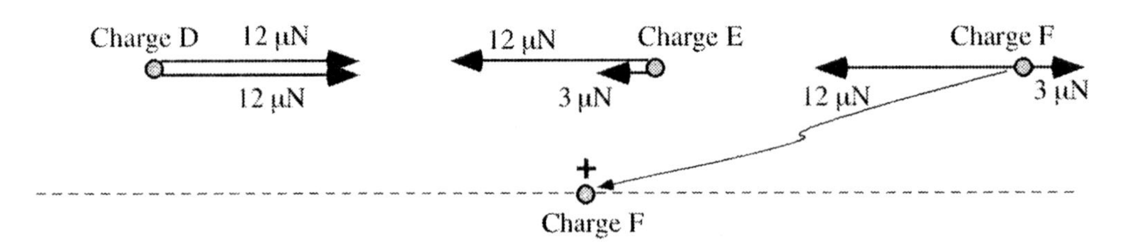

Case 3

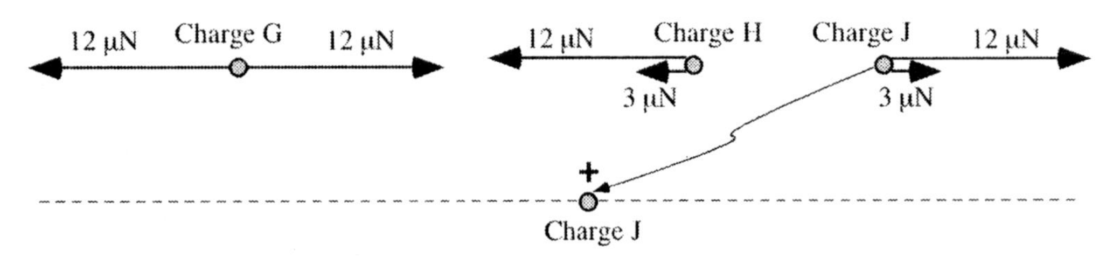

ET2-WBT11: FORCES ON THREE CHARGES IN TWO DIMENSIONS—CHARGE LOCATIONS

Shown at left below are three charged particles. Particles A and B have a charge $+2q$, and particle C has a charge $-q$. The diagrams show the electric force exerted on each particle due to the other two particles.

In the plane of the three charges, a coordinate system is chosen such that particle B lies at the origin. **On the grid, indicate the positions of particles A and C relative to particle B.**

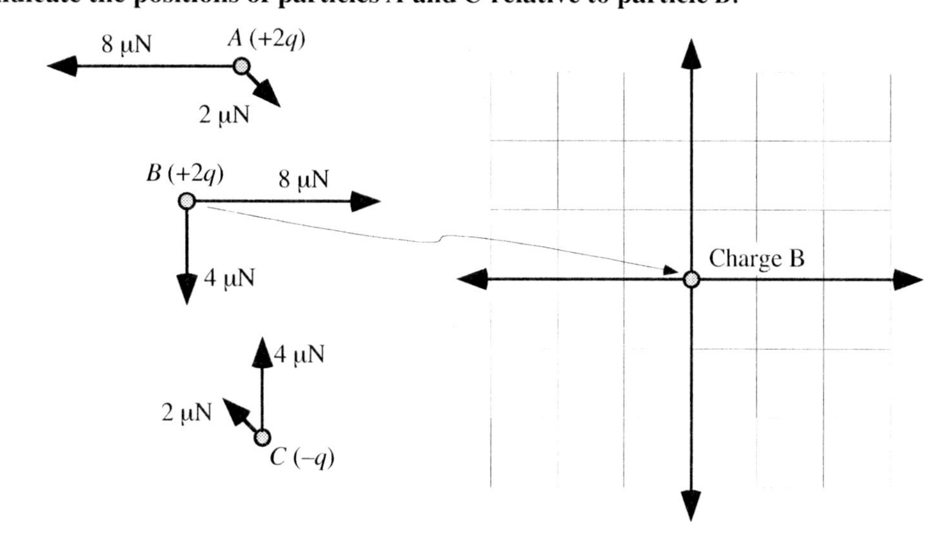

ET2-WBT12: POINT CHARGE INSIDE A SHELL—SHELL PROPERTIES

A point charge is inside of a shell, offset from the center of the shell. The electric field is measured at four labeled points outside the shell. The electric field has the same magnitude at points B and C, and the same magnitude at points A and D. The magnitude of the field is smaller at points A and D than it is at points B and C.

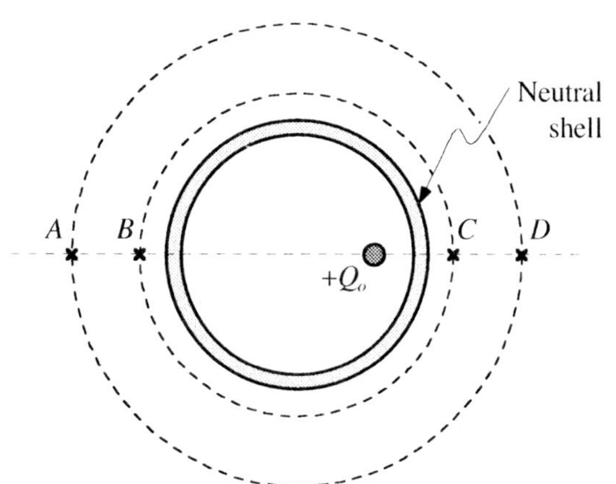

Based on the ranking for the electric field, determine whether the shell is a conductor or an insulator. Explain how you determined your answer. If you cannot determine this based on the ranking, explain why not.

MAGNETISM
RANKING TASKS (RT)

MT2-RT1: CHARGE WITHIN A UNIFORM MAGNETIC FIELD—MAGNETIC FORCE

The figures below show charged particles in an external magnetic field. All external magnetic fields are uniform and have the same strength.

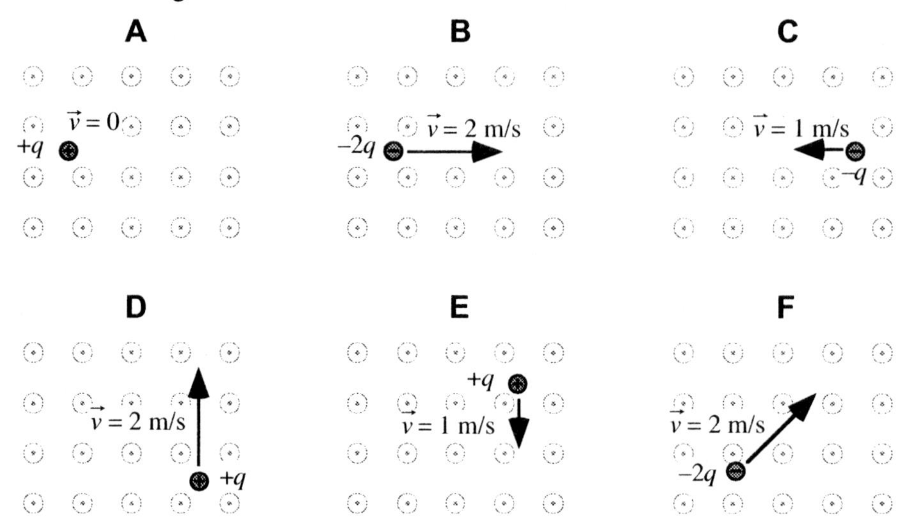

Rank the strength (magnitude) of the magnetic force on each charge.

Greatest 1 _______ 2 _______ 3 _______ 4 _______ 5 _______ 6 _______ Least

OR, the magnitude of the force is the same (but not zero) for all six situations. ______

OR, there is no magnetic force for all six situations. ______

OR, the ranking for the forces cannot be determined. ______

Carefully explain your reasoning.

How sure were you of your ranking? (circle one)

Basically Guessed					Sure				Very Sure
1	2	3	4	5	6	7	8	9	10

MT2-RT2: MOVING CHARGE PATH—DIRECTION AND STRENGTH OF THE MAGNETIC FIELD

Electrically charged particles of equal mass are moving through regions of space in which there may be magnetic fields. In each case, shown is the sign of the charge and a portion of the path the charge follows through the region. (These are top views looking down on horizontally moving charges.) All of the charges enter this region with the same initial velocity.

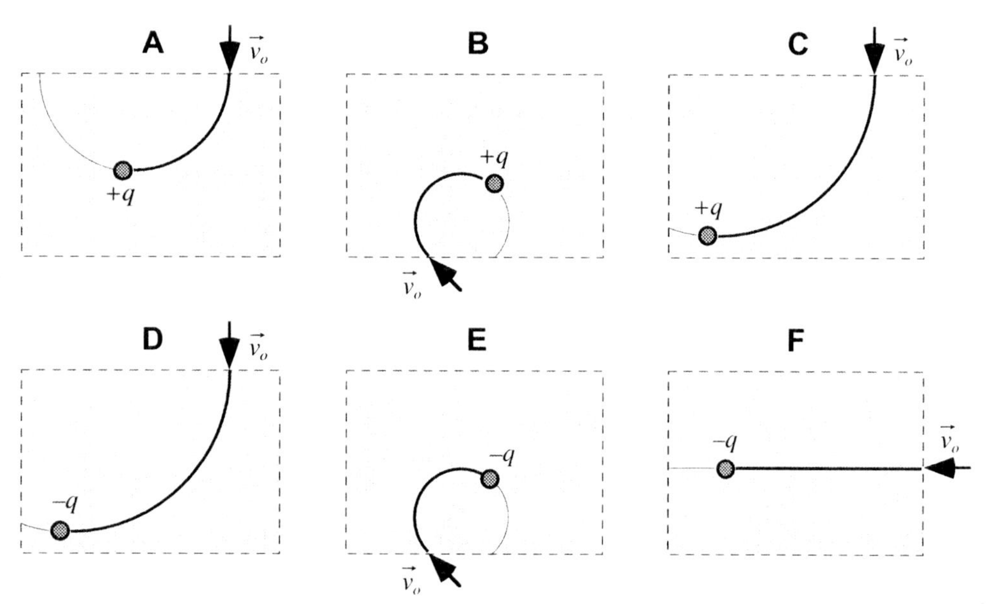

Rank the magnetic field in the region. Fields directed out of the page (considered positive) are ranked higher than fields directed into the page (considered negative).

Greatest out of page 1 ______ 2 ______ 3 ______ 4 ______ 5 ______ 6 ______ Greatest into page

OR, the magnetic field in all six of these cases has the same nonzero strength. _______

OR, the magnetic field is zero in all six of these cases. _______

OR, the ranking for the magnetic field cannot be determined. _______

Carefully explain your reasoning.

How sure were you of your ranking? (circle one)

Basically Guessed				Sure					Very Sure
1	2	3	4	5	6	7	8	9	10

MT2-RT3: PROTON IN MAGNETIC AND ELECTRIC FIELDS—ACCELERATION

A small region of space contains uniform magnetic and/or electric field(s). A proton enters this region from the left with a speed v_o.

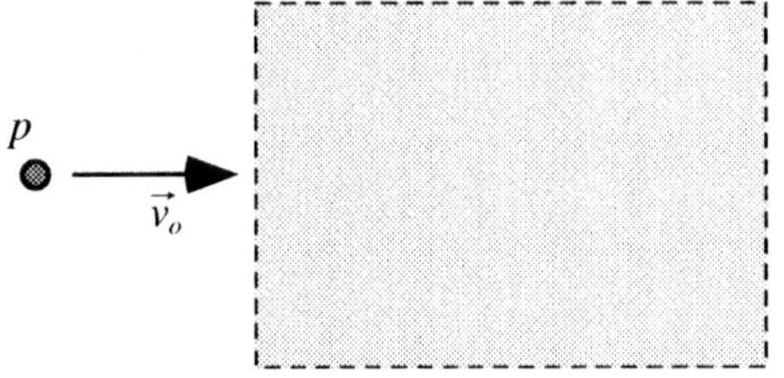

Region with uniform fields

The table below shows the direction of the electric and magnetic fields for seven cases labeled A-G.

Case	A	B	C	D	E	F	G
E-field direction	→	none	→	→	←	↑	↑
B-field direction	none	→	→	←	→	none	↑

Rank the magnitude of the acceleration of the proton just after it enters this region in these cases.

Greatest 1 _______ 2 _______ 3 _______ 4 _______ 5 _______ 6_______ 7 _______ Least

OR, the proton has the same but not zero acceleration for all seven cases. _____

OR, the proton has zero acceleration for all seven cases. _____

OR, it is not possible to determine the ranking for the acceleration for these cases. _____

Carefully explain your reasoning.

How sure were you of your ranking? (circle one)

Basically Guessed — Sure — Very Sure

1 2 3 4 5 6 7 8 9 10

MT3-RT1: MOVING CHARGE NEAR A STRAIGHT CURRENT–CARRYING WIRE—ACCELERATION

Six charged particles have been placed near identical current-carrying wires. These particles have the same mass, and have the same speed at the instant shown.

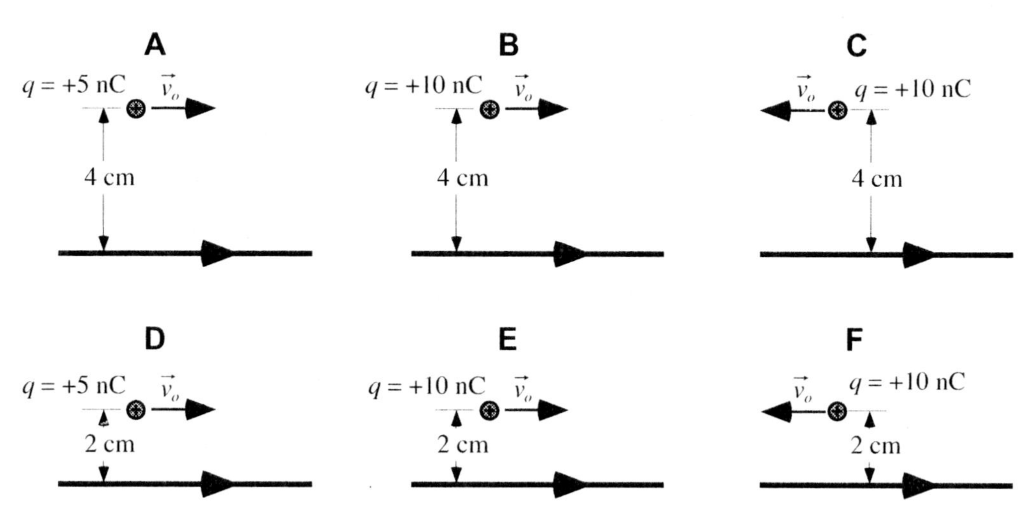

Rank the magnitude of the acceleration of each charge at the instant shown.

Greatest 1 _______ 2 _______ 3 _______ 4 _______ 5 _______ 6 _______ Least

OR, the acceleration is the same (but not zero) for all six situations. _______

OR, the acceleration is zero for all six situations. _______

OR, the ranking for the accelerations cannot be determined. _______

Carefully explain your reasoning.

How sure were you of your ranking? (circle one)

Basically Guessed				Sure					Very Sure
1	2	3	4	5	6	7	8	9	10

MT4-RT1: CURRENT–CARRYING WIRE IN A UNIFORM MAGNETIC FIELD—MAGNETIC FORCE

The figures below show identical current-carrying wire segments in identical uniform magnetic field regions. All the magnetic field regions are the same width and height.

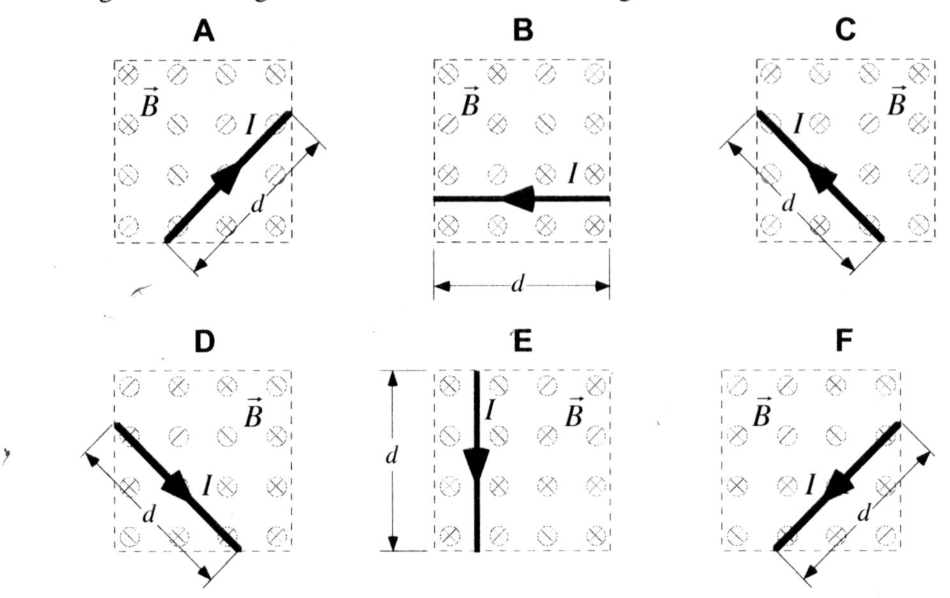

Rank the strength (magnitude) of the magnetic force on the wire segments.

Greatest 1 _______ 2 _______ 3 _______ 4 _______ 5 _______ 6 _______ Least

OR, the force is the same (but not zero) for all six situations. ______

OR, the force is zero for all six situations. ______

OR, the ranking for the forces cannot be determined. ______

Carefully explain your reasoning.

How sure were you of your ranking? (circle one)

Basically Guessed Sure Very Sure

1 2 3 4 5 6 7 8 9 10

MT6-RT1: STRAIGHT CURRENT–CARRYING WIRE—MAGNETIC FIELD

In each case shown below, a charged particle is placed at rest near a long current-carrying wire.

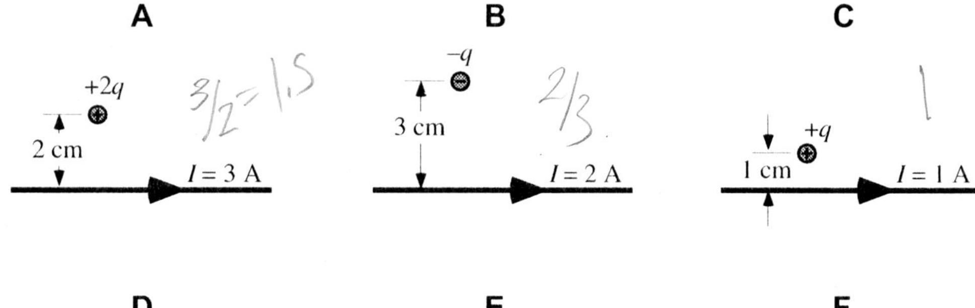

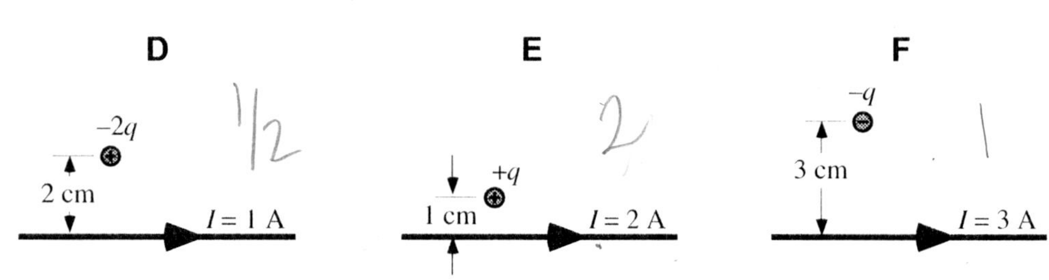

Rank the strength (magnitude) of the magnetic field at the location of these charges.

Greatest 1 ___E___ 2 ___A___ 3 ___C___ 4 ___F___ 5 ___B___ 6 ___D___ Least

OR, the magnetic field is the same (but not zero) for all six situations. ______

OR, the magnetic field is zero for all six situations. ______

OR, the ranking for the magnetic fields cannot be determined. ______

Carefully explain your reasoning.

How sure were you of your ranking? (circle one)

Basically Guessed — Sure — Very Sure

1 2 3 4 5 6 7 8 9 10

MT6-RT2: Three-Dimensional Locations near a Long Straight Wire—Magnetic Field

Shown below is a three-dimensional region that contains a long, straight, current-carrying wire in the *y-z* plane. Within that region are points located on the corners of two cubes as shown below. The small cube containing points *E*, *F*, and *D*, that also includes the origin, has edges of 1 cm length while the larger cube containing the other points, and the origin, has edges with a length of 3 cm. The current-carrying wire is aligned along the upper edge of the larger cube.

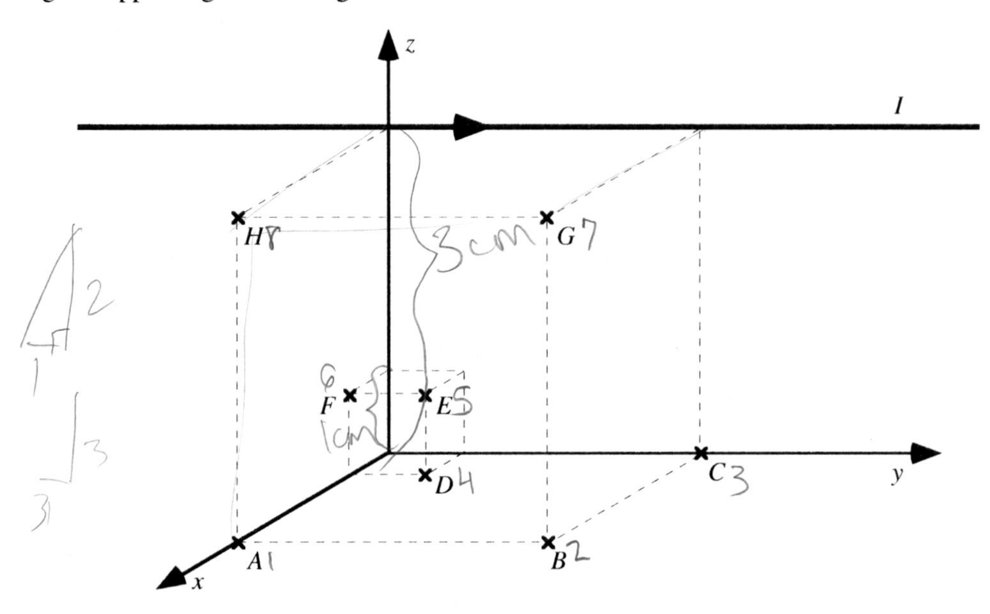

Rank the strength (magnitude) of the magnetic field at the labeled points.

Greatest 1 __E__ 2 __F__ 3 __G__ 4 __H__ 5 __C__ 6 __D__ 7 __A__ 8 __B__ Least

OR, the magnetic field is the same (but not zero) for all these points. _____

OR, the magnetic field is zero for all these points. _____

OR, the ranking for the magnetic field cannot be determined for all these points. _____

Carefully explain your reasoning.

How sure were you of your ranking? (circle one)

Basically Guessed | Sure | Very Sure

1 2 3 4 5 6 7 8 9 10

MT7-RT1: Current–Carrying Circular Loops—Magnetic Field

The figures below show current-carrying, circular loops. The current in each loop, the radius of the loops, and the number of turns making up the loop are given.

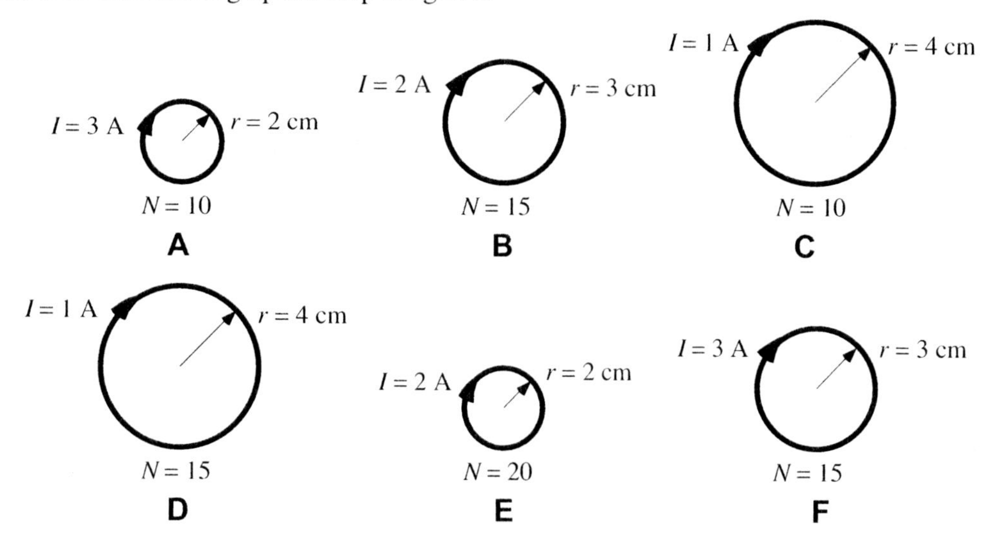

Rank the strength (magnitude) of the magnetic field at the center of the circular loops.

Greatest 1 _______ 2 _______ 3 _______ 4 _______ 5 _______ 6 _______ Least

OR, the magnetic field is the same (but not zero) for all six situations. ______

OR, the magnetic field is zero for all six situations. ______

OR, the ranking for the magnetic fields cannot be determined. ______

Carefully explain your reasoning.

How sure were you of your ranking? (circle one)

Basically Guessed				Sure				Very Sure
1	2	3	4	5 6	7	8	9	10

MT8-RT1: CURRENT–CARRYING STRAIGHT WIRES—MAGNETIC FIELD

Long straight wires that are perpendicular to the page are carrying electric currents into the page.

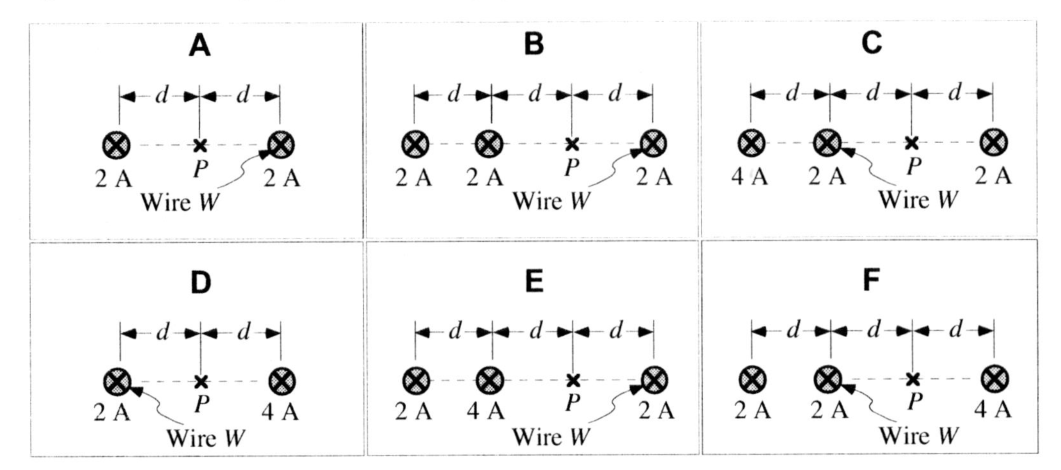

Rank the strength (magnitude) of the magnetic field at *P* due to wire *W*.

Greatest 1 _______ 2 _______ 3 _______ 4 _______ 5 _______ 6 _______ Least

OR, the magnetic field at *P* due to wire *W* is the same (but not zero) in all six cases. _______

OR, the magnetic field at *P* due to wire *W* is zero for all six situations. _______

OR, the ranking for the magnetic field at *P* due to wire *W* cannot be determined. _______

Carefully explain your reasoning.

How sure were you of your ranking? (circle one)

Basically Guessed				Sure					Very Sure
1	2	3	4	5	6	7	8	9	10

MT8-RT2: THREE PARALLEL CURRENT–CARRYING WIRES I—MAGNETIC FIELD

Shown below are six situations where three long straight parallel wires carry currents either into or out of the page. In each situation, point *P* is midway between two adjacent wires.

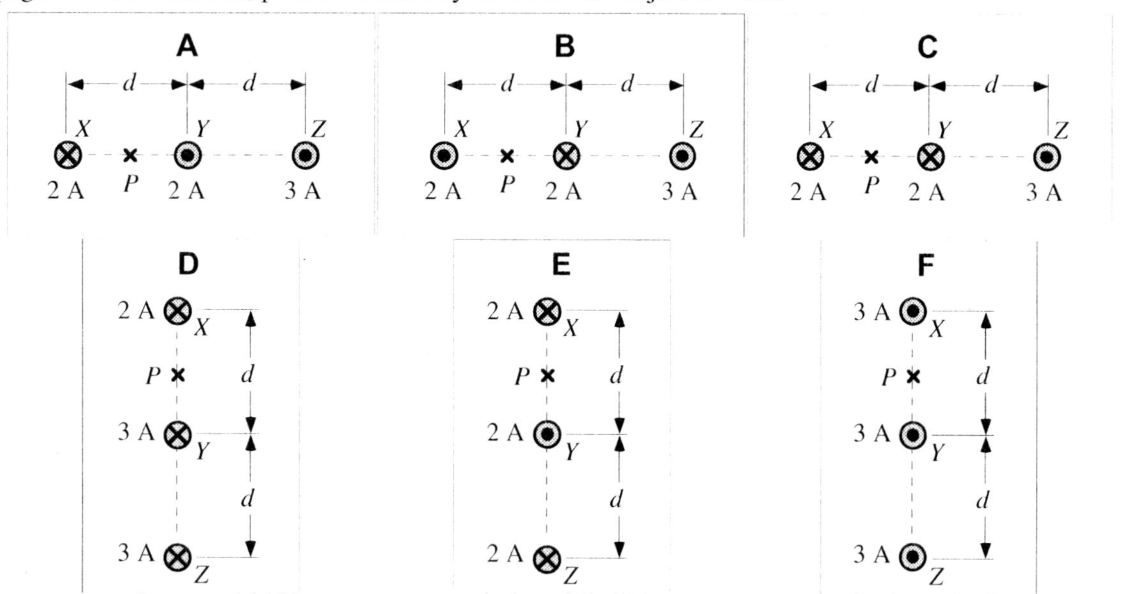

Rank the strength (magnitude) of the total magnetic field at point *P*.

Greatest 1 _______ 2 _______ 3 _______ 4 _______ 5 _______ 6 _______ Least

OR, the total magnetic field at *P* has the same (but not zero) strength for all six of these situations. _____

OR, the total magnetic field is zero at *P* for all six of these situations. _______

OR, the ranking for the magnetic fields cannot be determined. _______

Carefully explain your reasoning.

How sure were you of your ranking? (circle one)

Basically Guessed				Sure					Very Sure
1	2	3	4	5	6	7	8	9	10

MT8-RT3: Three Parallel Current–Carrying Wires II—Magnetic Field at Wire Y

Three parallel, long straight wires are carrying currents into or out of the page. The distance between wires *X* and *Y* is the same as the distance between wires *Y* and *Z*. The magnitude of the current in all wires is the same.

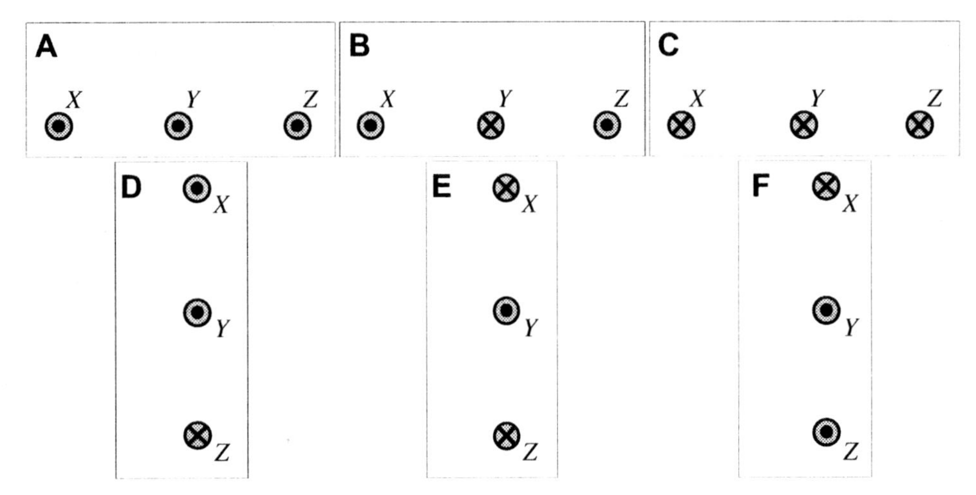

Rank the strength (magnitude) of the magnetic field acting on wire *Y*.

Greatest 1 _______ 2 _______ 3 _______ 4 _______ 5 _______ 6 _______ Least

OR, the magnetic field acting on wire *Y* is the same (but not zero) for all six of these situations. _____

OR, the magnetic field acting on wire *Y* is zero for all six situations. _____

OR, the ranking for the magnetic field acting on wire *Y* cannot be determined. _______

Carefully explain your reasoning.

How sure were you of your ranking? (circle one)

Basically Guessed				Sure					Very Sure
1	2	3	4	5	6	7	8	9	10

MT9-RT1: PARALLEL CURRENT–CARRYING WIRES I—MAGNETIC FORCE ON WIRE

In each case below, two or three very long straight wires are parallel to each other.

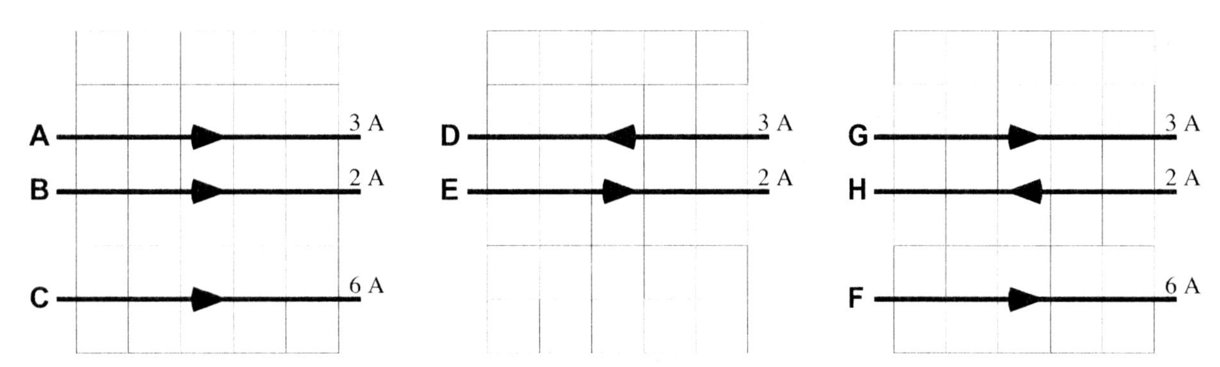

Rank the strength (magnitude) of the total magnetic force on each of the labeled wires.

Greatest 1 _____ 2 _____ 3 _____ 4 _____ 5 _____ 6 _____ 7 _____ 8 _____ Least

OR, the total magnetic force is the same (but not zero) for all of these wires. _____

OR, the magnetic force is zero for all eight wires. _____

OR, the ranking for the total magnetic force cannot be determined. _____

Carefully explain your reasoning.

How sure were you of your ranking? (circle one)

Basically Guessed				Sure					Very Sure
1	2	3	4	5	6	7	8	9	10

MT10-RT1: MOVING CHARGE IN A UNIFORM MAGNETIC FIELD—CHANGE IN KINETIC ENERGY

Identical positively charged particles are moving through uniform external magnetic fields. The speed of each particle as it entered the field is given and the strength of the magnetic field is indicated in each diagram. All particles remain within the magnetic field for at least three seconds.

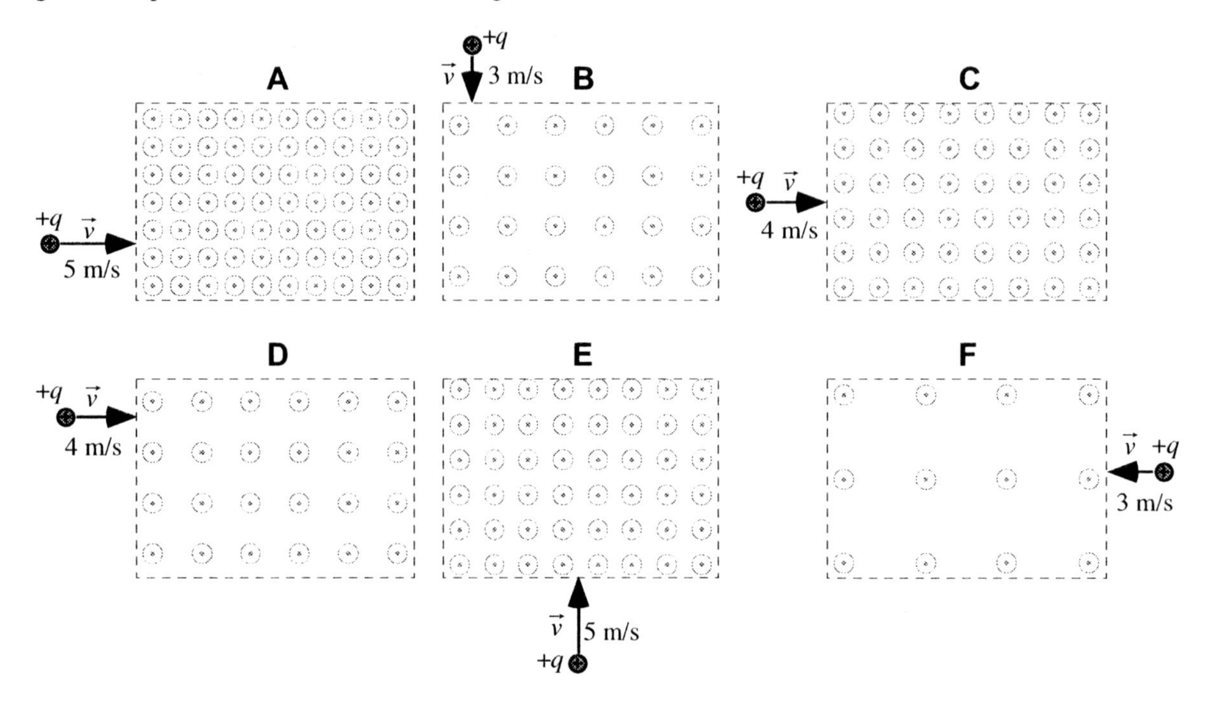

Rank the change in kinetic energy of the particles during their first three seconds in the fields.

Greatest 1 _______ 2 _______ 3 _______ 4 _______ 5 _______ 6 _______ Least

OR, the change in kinetic energy will be the same (but not zero) for all of these situations. _______

OR, there will be NO change in kinetic energy for any of these situations. _______

OR, the ranking for the change in kinetic energy cannot be determined. _______

Carefully explain your reasoning.

How sure were you of your ranking? (circle one)

Basically Guessed				Sure					Very Sure
1	2	3	4	5	6	7	8	9	10

MT11-RT1: Moving Rectangular Loops in Uniform Magnetic Fields—Magnetic Flux

Eight identical rectangular wire loops are moving to the right in the vicinity of a region in which there is a uniform magnetic field coming out of the page. The rectangular loops are being pushed to the right at a constant speed. The loops have various portions (25%, 50%, 75%, or 100%) within the field as shown.

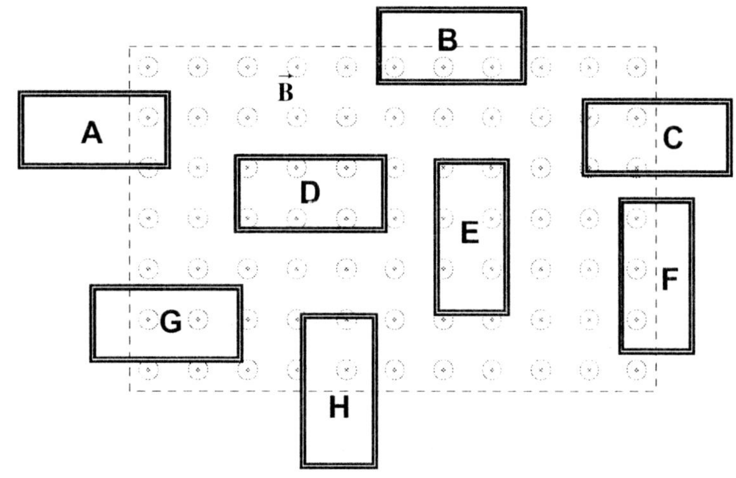

All loops moving to the right

Rank the magnetic flux through the rectangular loops at the positions shown. Assume there is no interaction between the loops.

Greatest 1 _____ 2 _____ 3 _____ 4 ______ 5 _____ 6 _____ 7 _____ 8 _____ Least

OR, all the loops have the same (but not zero) magnetic flux through them. _______

OR, there is no magnetic flux through any of the loops. _______

OR, the ranking for the magnetic flux cannot be determined. _______

Carefully explain your reasoning.

How sure were you of your ranking? (circle one)

Basically Guessed					Sure				Very Sure
1	2	3	4	5	6	7	8	9	10

MT11-RT2: Moving Rectangular Loops in Uniform Magnetic Fields—Magnetic Flux

Eight identical rectangular wire loops are moving to the right in the vicinity of a region in which there is a uniform magnetic field coming out of the page. The rectangular loops are being pushed to the right at various constant speeds as indicated in the diagram. The loops have various portions (25%, 50%, 75%, or 100%) within the field as shown.

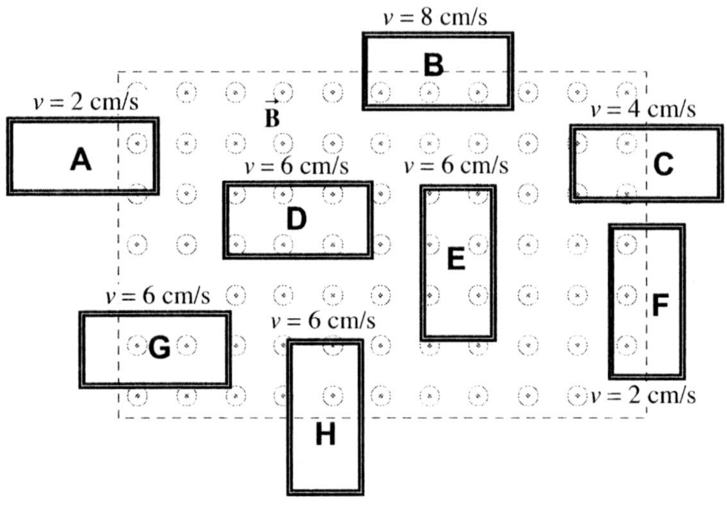

All loops moving to the right

Rank the magnitude of the magnetic flux through the rectangular loops at the positions shown.
Assume there is no interaction between the loops.

Greatest 1 ______ 2 ______ 3 ______ 4 ______ 5 ______ 6 ______ 7 ______ 8 ______ Least

OR, all the loops have the same (but not zero) magnetic flux. ______

OR, there is no magnetic flux through any of the loops. ______

OR, the ranking for the magnetic flux cannot be determined. ______

Carefully explain your reasoning.

How sure were you of your ranking? (circle one)

Basically Guessed				Sure					Very Sure
1	2	3	4	5	6	7	8	9	10

MT11-RT3: MOVING RECTANGULAR LOOPS IN UNIFORM MAGNETIC FIELDS—MAGNETIC FLUX CHANGE

Eight identical rectangular wire loops are moving to the right in the vicinity of a region in which there is a uniform magnetic field coming out of the page. The rectangular loops are being pushed to the right all at the same constant speed. The loops have various portions (25%, 50%, 75%, or 100%) within the field as shown.

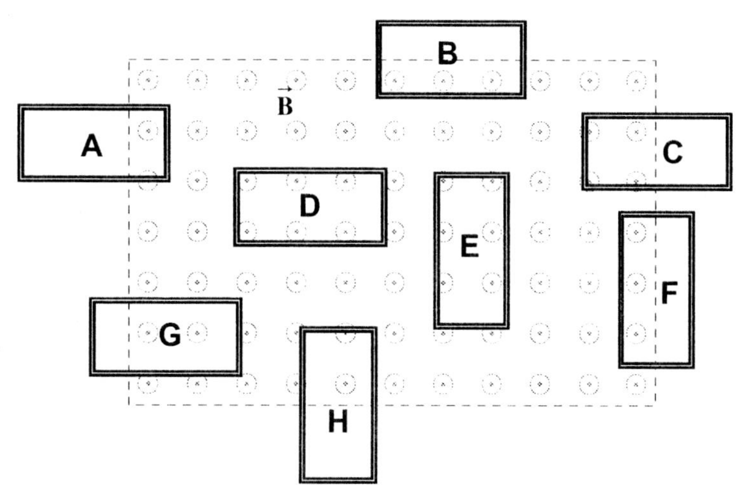

All loops moving to the right

Rank the magnitude of the rate of change of magnetic flux through the rectangular loops at the positions shown. Assume there is no interaction between the loops.

Greatest 1 _____ 2 _____ 3 _____ 4 _____ 5 _____ 6 _____ 7 _____ 8 _____ Least

OR, all the loops have the same (but not zero) change in magnetic flux through them. _______

OR, there is no change in magnetic flux through any of the loops. _______

OR, the ranking for the change in magnetic flux cannot be determined. _______

Carefully explain your reasoning.

How sure were you of your ranking? (circle one)

Basically Guessed				Sure					Very Sure
1	2	3	4	5	6	7	8	9	10

MT11-RT4: Moving Rectangular Loops in Uniform Magnetic Fields—Magnetic Flux Change

Eight identical rectangular wire loops are moving to the right in the vicinity of a region in which there is a uniform magnetic field coming out of the page. The rectangular loops are being pushed to the right at various constant speeds as indicated in the diagram. The loops have various portions (25%, 50%, 75%, or 100%) within the field as shown. The rectangular loops are all 5 cm by 10 cm.

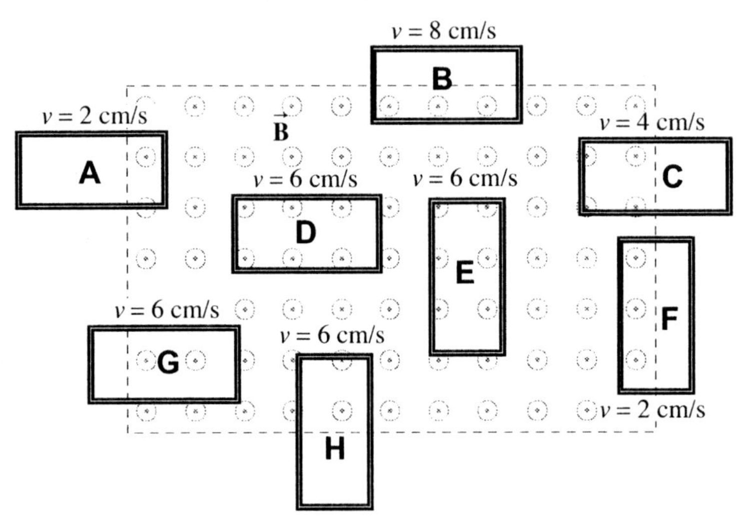

All loops moving to the right

Rank the magnitude of the rate of change of magnetic flux through the rectangular loops at the positions shown. Assume there is no interaction between the loops.

Greatest 1 _____ 2 _____ 3 _____ 4 _____ 5 _____ 6 _____ 7 _____ 8 _____ Least

OR, all the loops have the same (but not zero) rate of change of magnetic flux. _______

OR, the rate of change of magnetic flux is zero for all of the loops. _______

OR, the ranking for the rate of change of magnetic flux cannot be determined. _______

Carefully explain your reasoning.

How sure were you of your ranking? (circle one)

Basically Guessed				Sure					Very Sure
1	2	3	4	5	6	7	8	9	10

MT12-RT1: Moving Rectangular Loops in Uniform Magnetic Fields—Voltage

Six identical rectangular wire loops are moving to the right at the same speed in the vicinity of a region with uniform magnetic fields. The strengths of the magnetic fields vary as indicated. All loops contain a one ohm resistor that is not shown.

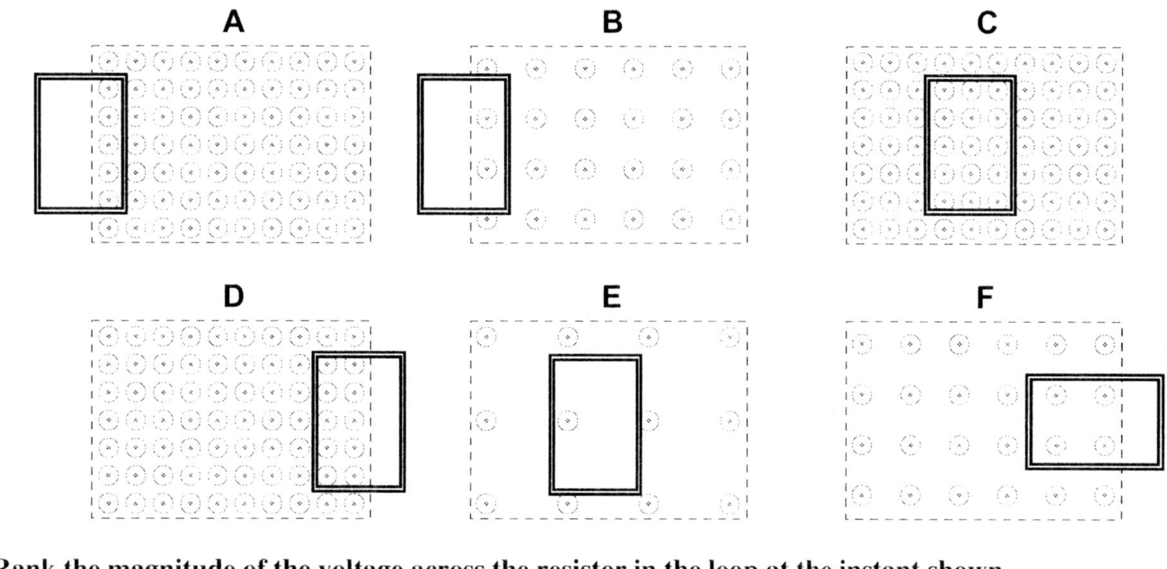

Rank the magnitude of the voltage across the resistor in the loop at the instant shown.

Greatest 1 _______ 2 _______ 3 _______ 4 _______ 5 _______ 6 _______ Least

OR, the voltage is the same (but not zero) for all situations. _______

OR, all six of these situations have zero voltage. _______

OR, the ranking for the voltage cannot be determined. _______

Carefully explain your reasoning.

How sure were you of your ranking? (circle one)

Basically Guessed				Sure					Very Sure
1	2	3	4	5	6	7	8	9	10

MT13-RT1: Moving Rectangular Loops in Uniform Magnetic Fields—Current

Eight identical rectangular wire loops are moving to the right in the vicinity of a region in which there is a uniform magnetic field coming out of the page. The rectangular loops are being pushed to the right at various constant speeds. The loops have various portions (25%, 50%, 75%, or 100%) within the field as shown. The rectangular loops are all 5 cm by 10 cm.

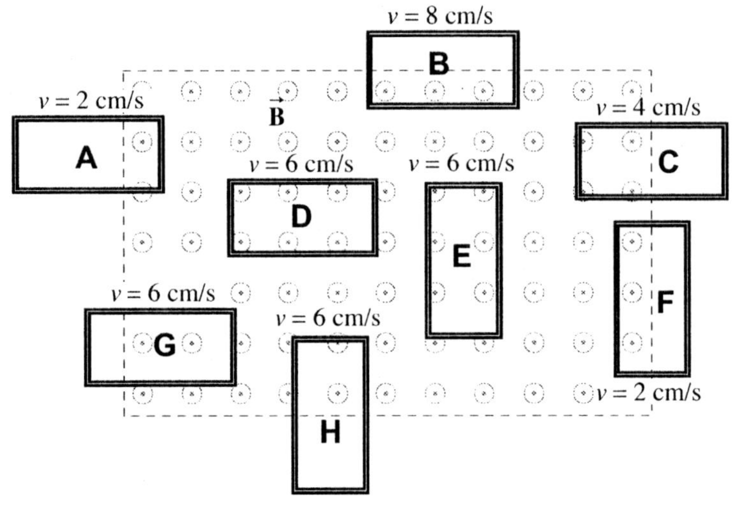

All loops moving to the right

Rank the magnitude of the current in the rectangular loops at the time and positions shown. Assume there is no effect or interaction between the loops.

Greatest 1 _____ 2 _____ 3 _____ 4 _____ 5 _____ 6 _____ 7 _____ 8 _____ Least

OR, all the loops have the same (but not zero) current. _______

OR, there is no current in any of the loops. _______

OR, the ranking for the current cannot be determined. _______

Carefully explain your reasoning.

How sure were you of your ranking? (circle one)

Basically Guessed — Sure — Very Sure

1 2 3 4 5 6 7 8 9 10

MT13-RT2: Changing Current—Bulb Brightness

In each case shown below, a long straight wire is sitting next to a circular wire loop that has a small light bulb in it. The currents in the long wires are changing at uniform rates. The initial current, the final current, and the time interval during which the change occurs are given in each figure. The bulbs and wire loops are all identical, and the straight wires are all the same distance from the wire loops.

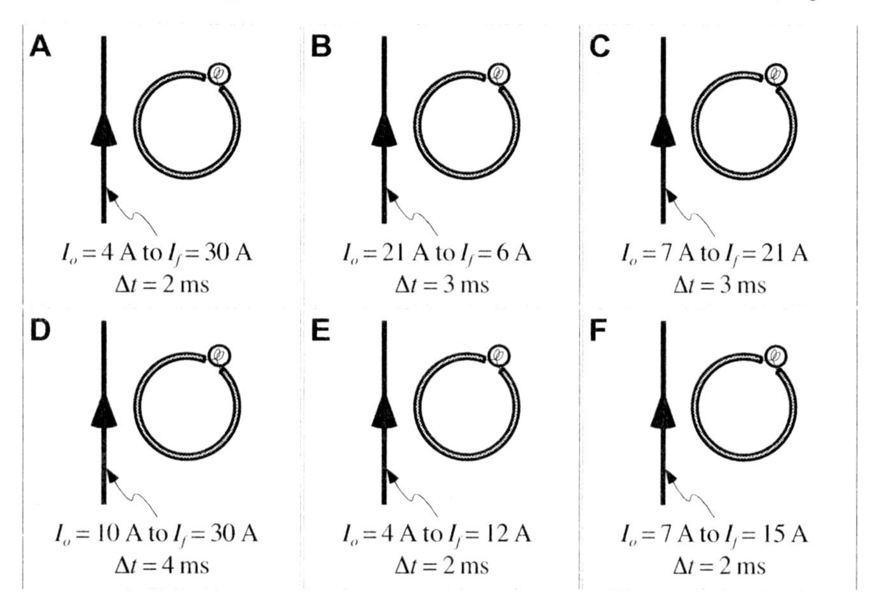

Rank the bulb brightness during the time that the current changes.

Greatest 1 _______ 2 _______ 3 _______ 4 _______ 5 _______ 6 _______ Least

OR, the bulbs in all six of these cases will have the same brightness. _______

OR, none of the bulbs will light for these six cases. _______

OR, the ranking for the brightness cannot be determined. _______

Carefully explain your reasoning.

How sure were you of your ranking? (circle one)

Basically Guessed | Sure | Very Sure

1 2 3 4 5 6 7 8 9 10

MT14-RT1: WIRE ON A LOOP MOVING IN A MAGNETIC FIELD—MAGNETIC FIELD

Each vertical conducting bar forms a complete rectangular loop with the conductor it is sliding on. The vertical bars are moving to the right at different constant speeds through a region of external uniform magnetic field.

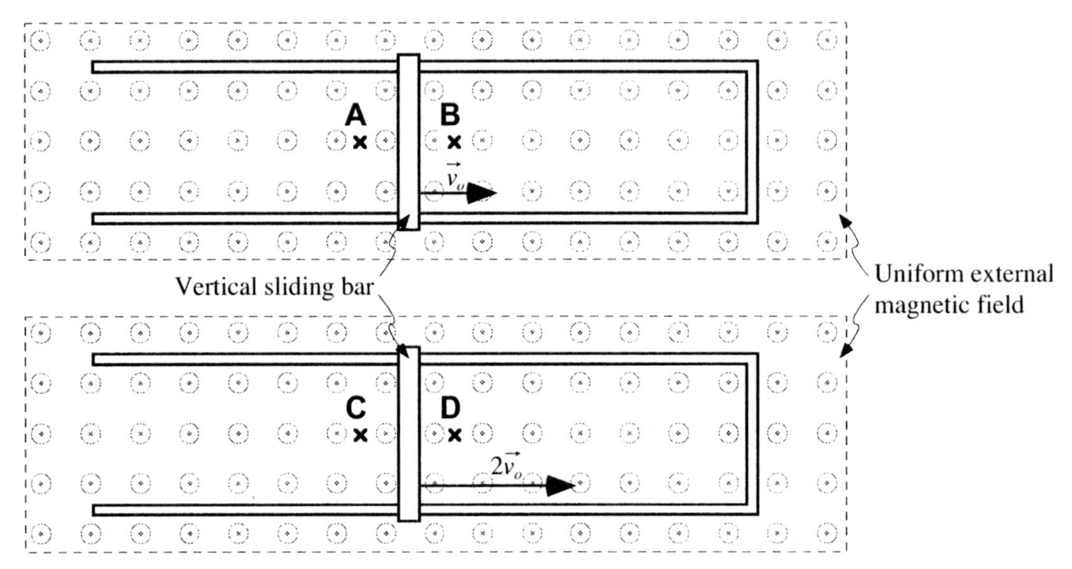

At the instant shown, rank the strength (magnitude) of the total magnetic field at the labeled positions.

Greatest 1 ______ 2 ______ 3 ______ 4 ______ Least

OR, all the points have the same (but not zero) strength magnetic field. ______

OR, the ranking for the magnetic field strength cannot be determined. ______

Carefully explain your reasoning.

How sure were you of your ranking? (circle one)

Basically Guessed					Sure				Very Sure
1	2	3	4	5	6	7	8	9	10

MT14-RT2: Loop Moving into a Uniform Magnetic Field—Magnetic Field

A rectangular wire loop is moving to the right at a constant speed into a region in which there is a uniform external magnetic field coming out of the page. At the instant shown, the loop is halfway into the magnetic field region. Points *B*, *C*, *D*, and *E* are the same distance from the loop while *A* and *F* are the same larger distance from the loop.

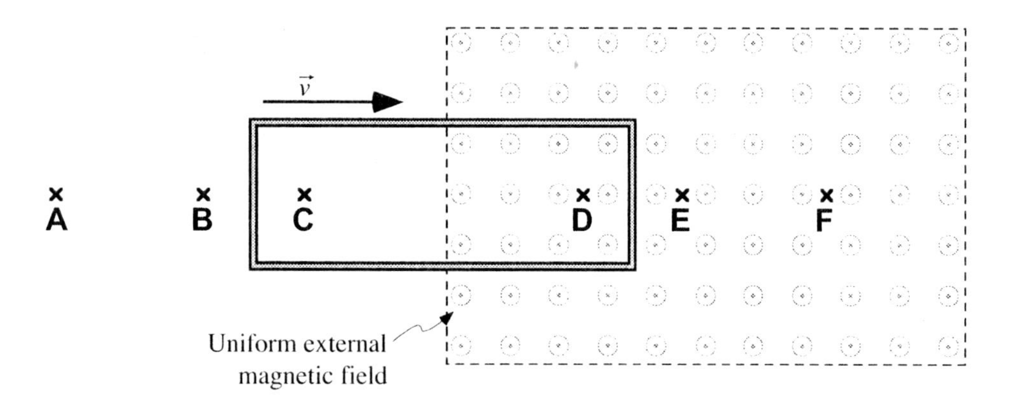

Rank the strength (magnitude) of the total magnetic field at the labeled points.

Greatest 1 _______ 2 _______ 3 _______ 4 _______ 5 _______ 6 _______ Least

OR, the magnetic field is the same (but not zero) at all these points. _______

OR, the magnetic field is zero at all these points. _______

OR, the ranking for the magnetic field cannot be determined. _______

Carefully explain your reasoning.

How sure were you of your ranking? (circle one)

Basically Guessed				Sure				Very Sure
1	2	3	4	5 6	7	8	9	10

MT14-RT3: LOOPS AND UNIFORM MAGNETIC FIELDS—MAGNETIC FIELD

Eight identical rectangular wire loops are in a region in which there is a uniform magnetic field coming out of the page. Six of loops are being pushed to the right at various constant speeds and two of the loops are not moving as indicated. The loops have various portions (25%, 50%, 75%, or 100%) within the field as shown.

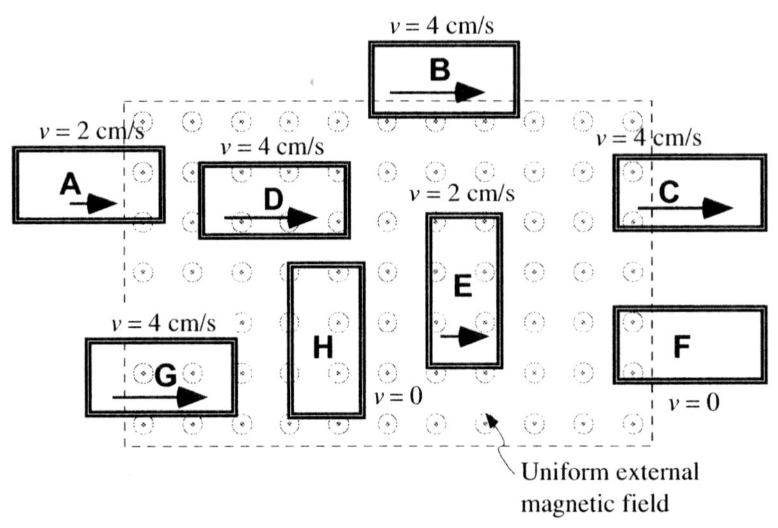

Rank the strength (magnitude) of the total magnetic field at the center of the rectangular loops. Assume there is no interaction between the loops.

Greatest 1 _____ 2 _____ 3 _____ 4 _____ 5 _____ 6 _____ 7 _____ 8 _____ Least

OR, all the loops have the same (but not zero) magnetic field at the center. _______

OR, the ranking for the magnetic field cannot be determined. _______

Carefully explain your reasoning.

How sure were you of your ranking? (circle one)

Basically Guessed — Sure — Very Sure

1 2 3 4 5 6 7 8 9 10

MT14-RT4: WIRE ON A LOOP MOVING IN A MAGNETIC FIELD—INDUCED MAGNETIC FIELD

Each vertical conducting bar forms a complete rectangular loop with the conductor it is sliding on. The vertical bars are moving to the right at different constant speeds through a region of an external uniform magnetic field.

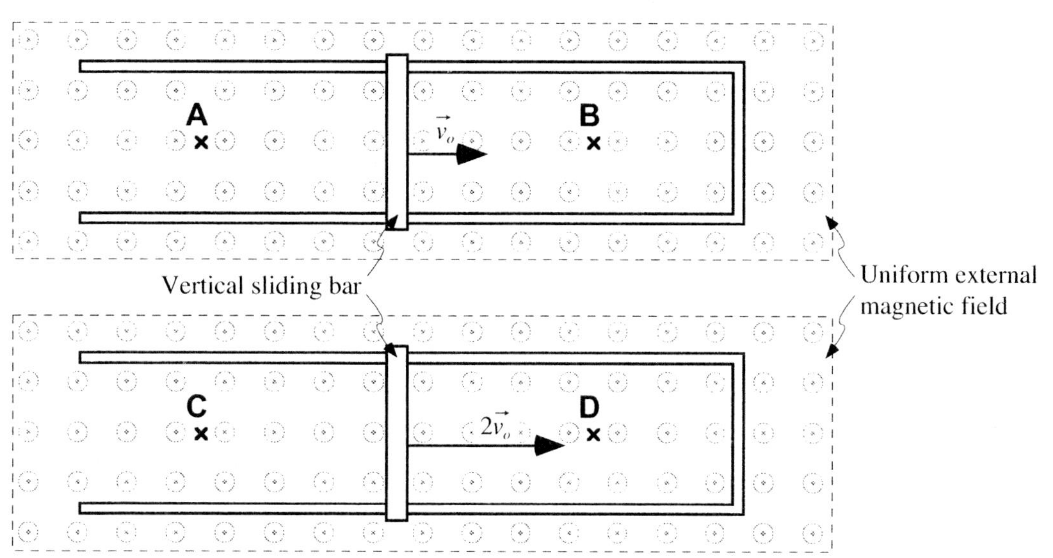

Rank the induced magnetic field at the four labeled points. Assume fields out of the page to be "positive" and fields into the page to be "negative."

Greatest out of the page 1 ______ 2 ______ 3 ______ 4 ______ Greatest into the page

OR, all the points have the same (but not zero) induced magnetic field. ______

OR, the induced magnetic field at all these points is zero. ______

OR, the ranking for the induced magnetic field cannot be determined. ______

Carefully explain your reasoning.

How sure were you of your ranking? (circle one)

Basically Guessed				Sure					Very Sure
1	2	3	4	5	6	7	8	9	10

MT14-RT5: LOOPS AND UNIFORM MAGNETIC FIELD—INDUCED MAGNETIC FIELD

Eight identical rectangular wire loops are in a region in which there is a uniform magnetic field coming out of the page. Six of loops are being pushed to the right at various constant speeds and two of the loops are not moving as indicated. The loops have different locations with various portions (25%, 50%, 75%, or 100%) within the field as shown.

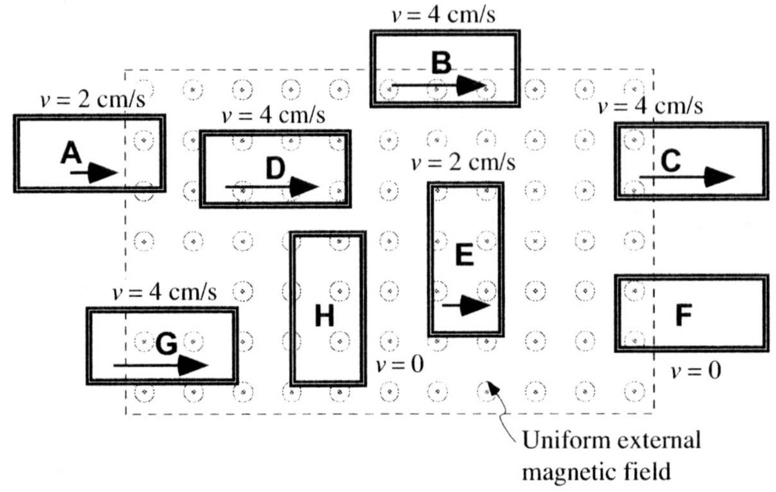

Rank the induced magnetic field at the center of the rectangular loops at the positions shown. Assume fields out of the page to be "positive" and fields into the page to be "negative." Also assume there is no interaction between the loops.

Greatest out of the page 1____ 2_____ 3____ 4____ 5____ 6____ 7____ 8____ Greatest into the page

OR, all the loops have the same (but not zero) induced magnetic field at the center of the loops. _______

OR, the induced magnetic field at all these points is zero. ______

OR, the ranking for the induced magnetic field cannot be determined. _______

Carefully explain your reasoning.

How sure were you of your ranking? (circle one)

Basically Guessed				Sure					Very Sure
1	2	3	4	5	6	7	8	9	10

MT15-RT1: Wire Loops and Moving Magnets—Loop Motion

Six permanent magnets are moving toward the circular wire loops that are fixed in place. All of the wire loops are the same radius and are made of the same wire, but some of them have gaps. The magnets are all identical, but they are approaching the loops at different speeds and with different orientations.

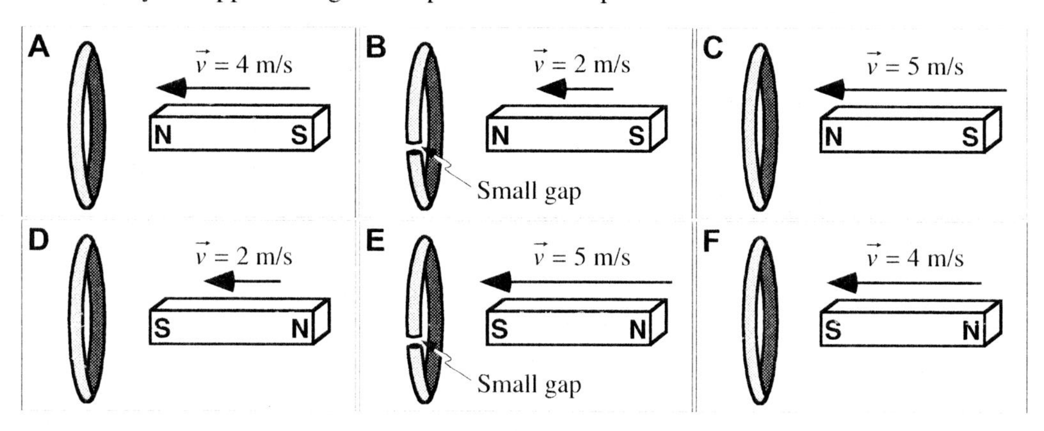

Rank how strongly the loops are repelled by the approaching magnet when the magnets are 3 cm from the loops.

Greatest 1 ______ 2 ________ 3 _______ 4 ________ 5 _______ 6 ________ Least

OR, all six of these loops will experience the same (but not zero) repulsion. ________

OR, there is no repulsion in any of these cases. ________

OR, the ranking for the repulsion cannot be determined. ________

Carefully explain your reasoning.

How sure were you of your ranking? (circle one)

Basically Guessed				Sure					Very Sure
1	2	3	4	5	6	7	8	9	10

COMPARISON TASKS (CT)

MT11-CT1: MOVING RECTANGULAR LOOPS IN UNIFORM MAGNETIC FIELDS—MAGNETIC FLUX

In each case below, a rectangular loop of wire is moving to the right at 4 m/s into a region with a uniform magnetic field pointing into the page. In Case B, the field is four times as strong as in Case A. The rectangular loops have different widths and heights, but they are made from identical wires. At the instant shown, both loops are halfway into the uniform magnetic field regions.

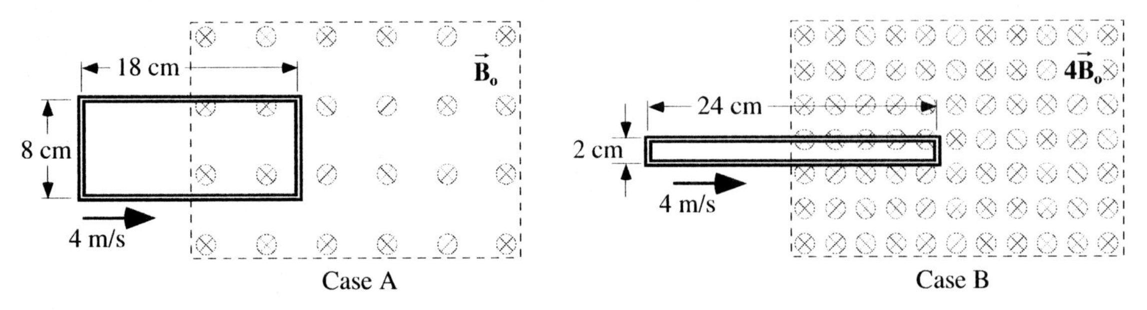

Is the magnetic flux in the loop in Case A *greater than*, *less than*, or *equal to* the magnetic flux in the loop in Case B at the instant shown? Explain.

MT11-CT2: MOVING RECTANGULAR LOOPS IN UNIFORM MAGNETIC FIELDS—MAGNETIC FLUX CHANGE

In each case below, a rectangular loop of wire is moving to the right at 4 m/s into a region with a uniform magnetic field pointing into the page. In Case B, the field is four times as strong as in Case A. The rectangular loops have different widths and heights, but they are made from identical wires. At the instant shown, both loops are halfway into the uniform magnetic field regions.

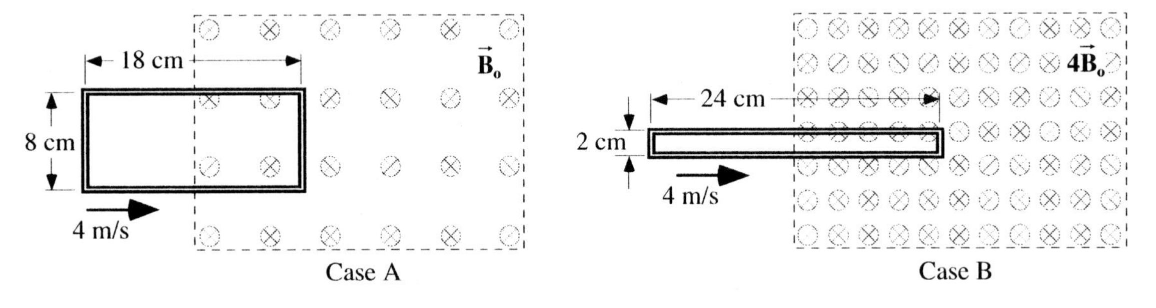

Is the rate of change in magnetic flux in the loop in Case A *greater than*, *less than*, or *equal to* the rate of change in magnetic flux in the loop in Case B at the instant shown? Explain.

MT13-CT1: MOVING RECTANGULAR LOOPS IN UNIFORM MAGNETIC FIELDS—CURRENT

In each case below, a rectangular loop of wire is moving to the right at 4 m/s into a region with a uniform magnetic field pointing into the page. In Case B, the field is four times as strong as in Case A. The rectangular loops have different widths and heights, but they are made from identical wires. At the instant shown, both loops are halfway into the uniform magnetic field regions. There is an induced current in the left loop of 12 mA.

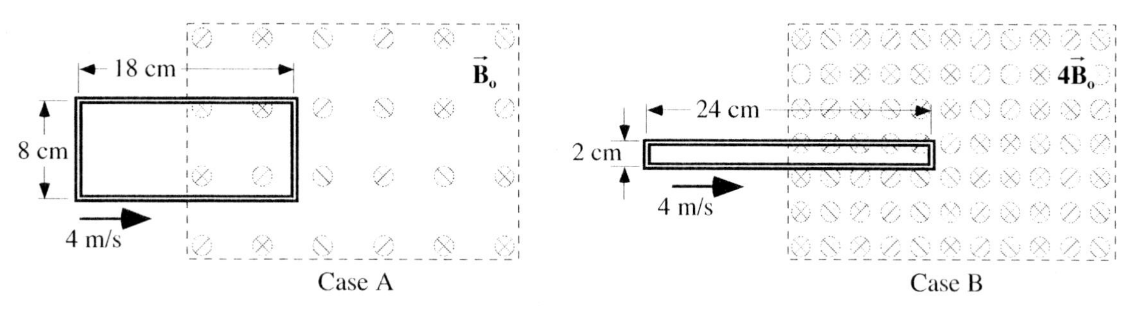

Case A Case B

Is the induced current in the loop in Case A *greater than*, *less than*, or *equal to* the induced current in the loop in Case B at the instant shown? Explain.

MT13-CT2: MOVING RECTANGULAR LOOPS IN UNIFORM MAGNETIC FIELDS—CURRENT

In each case below, a rectangular loop of wire is moving to the right at the same velocity into a region with a uniform external magnetic field pointing into the page. The magnetic field is the same in both cases. The rectangular loops have different widths but have the same height. The loops are made from copper wires that have the same cross-section. At the instant shown, the leading edge of both loops is the same distance into the uniform magnetic field region.

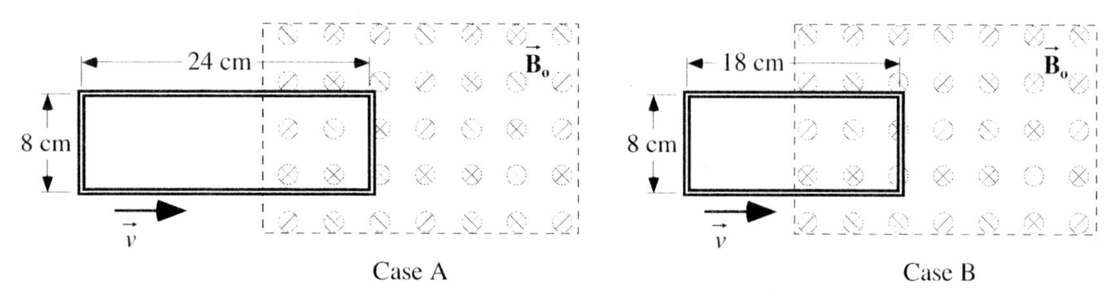

Case A Case B

Is the induced current in the loop in Case A *greater than*, *less than*, or *equal to* the induced current in the loop in Case B at the instant shown? Explain.

QUALITATIVE REASONING TASKS (QRT)

MT1-QRT1: ELECTRIC CHARGE NEAR A BAR MAGNET—FORCE DIRECTION

An electrically charged particle is placed near one pole of a permanent magnet. All of the charges are at rest and are the same distance from the magnet. The strength of the magnetic field due to the bar magnet at the location of the particle is given.

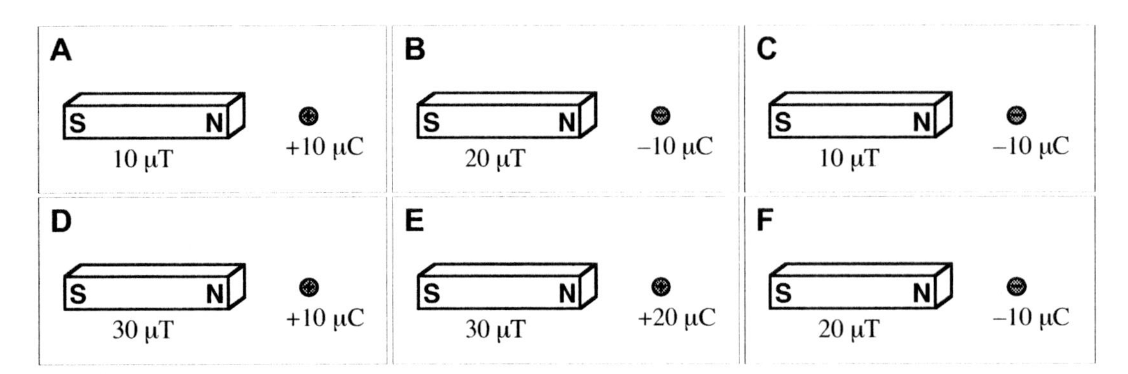

For these situations, draw the direction of the initial force on the magnet and on the charged particle in the chart below.

	Direction of force on the charged particle	Direction of force on the magnet
Case A		
Case B		
Case C		
Case D		
Case E		
Case F		

Carefully explain your reasoning.

MT1-QRT2: Charge near a Circular Current Loop—Magnetic Force Direction

An electrically charged particle is placed at rest near a circular current-carrying loop of wire, along the center line of the loop. All of the charges have the same magnitude and the currents in all the loops are in the same direction.

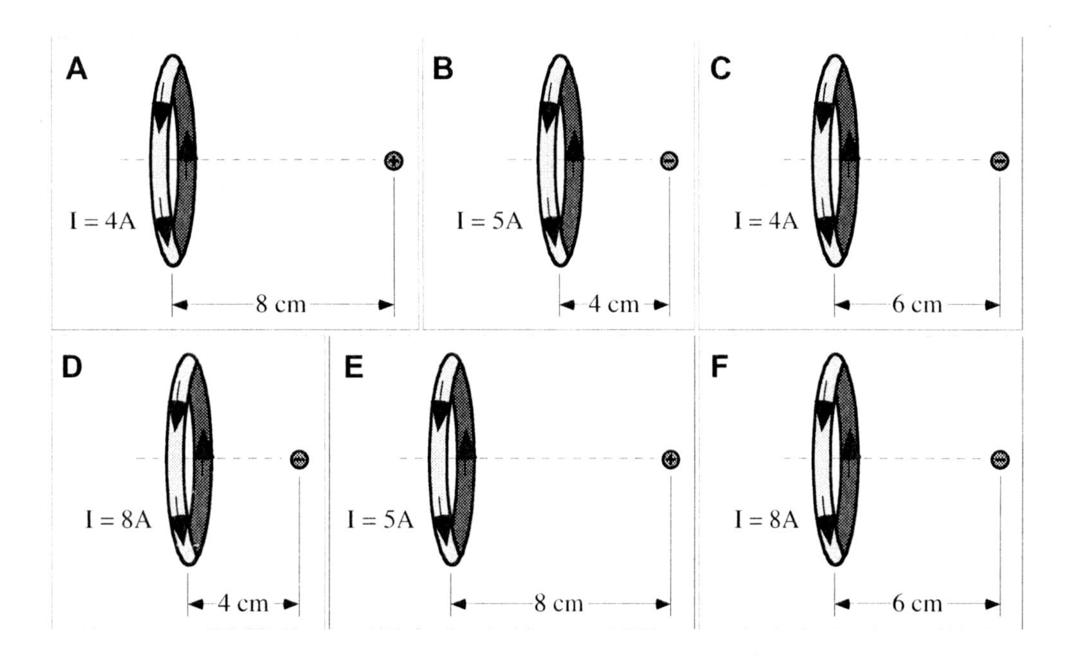

For these situations, draw the direction of the magnetic force exerted on the charged particle and on the current loop in the chart below.

	Direction of the magnetic force on the charged particle	Direction of the magnetic force on the current loop		Direction of the magnetic force on the charged particle	Direction of the magnetic force on the current loop
Case A			Case D		
Case B			Case E		
Case C			Case F		

Carefully explain your reasoning.

MT2-QRT1: CHARGED PARTICLE AND A UNIFORM MAGNETIC FIELD—PATH

The figure below shows the path of a negatively charged particle in a region of a uniform magnetic field.

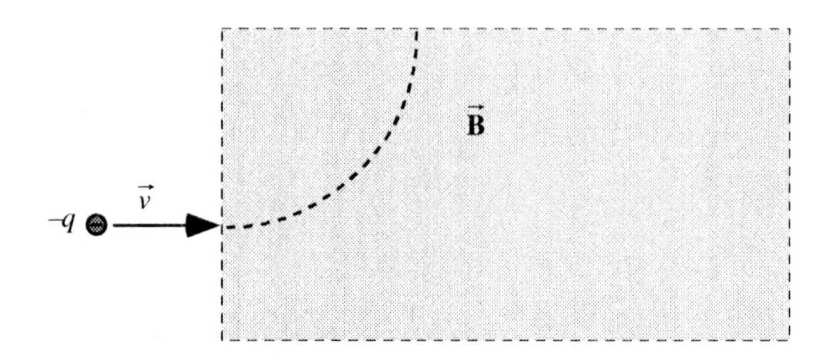

1) What is the direction of the magnetic field in the shaded region?

2) If we double the speed of the particle, how will the path change?

3) If we double the magnitude of the uniform magnetic field, how will the path change?

4) If we replace the original particle with a negative particle of twice the charge and the same mass, how will the path change?

5) If we replace the original particle with a positive particle of the same mass and same magnitude charge as the original negative charge, how will the path change?

6) If we replace the original particle with a negative particle of twice the mass and the same charge, how will the path change?

MT3-QRT1: Moving Charge near a Straight Current–Carrying Wire—Acceleration

At the instant shown, a particle with charge of +7 nC is moving at 3 m/s parallel to a long straight wire that has a current of 8 A.

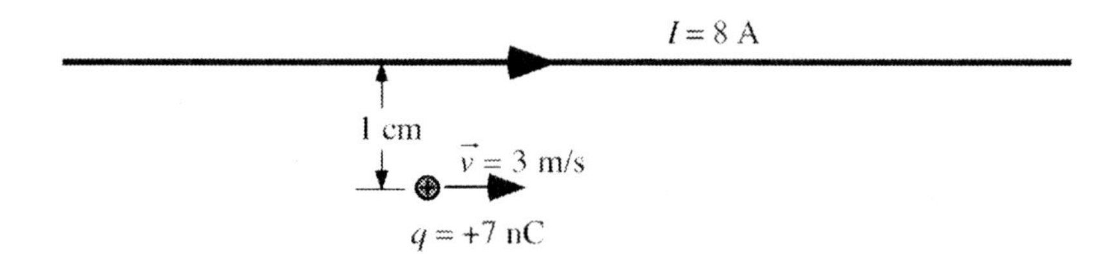

1) What is the direction of acceleration of the charged particle?

2) If we double the charge on the particle, what will happen to the acceleration?

3) If we replace the charge with a negative charge of the same mass and same magnitude charge as the original charge, what will happen to the acceleration?

4) If we double the distance from the wire to the particle, what will happen to the acceleration?

5) If we double the mass of the particle, what will happen to the acceleration?

6) If we double the velocity of the particle, what will happen to the acceleration?

7) If we reduce the magnitude of the current, what will happen to the acceleration?

8) If we reverse the direction of the current, what will happen to the acceleration?

MT4-QRT1: CURRENT–CARRYING WIRE IN A UNIFORM MAGNETIC FIELD—MAGNETIC FORCE

The figure below shows a current-carrying straight wire segment (with connecting wires up out of the paper that are not shown) in a uniform magnetic field directed into the paper.

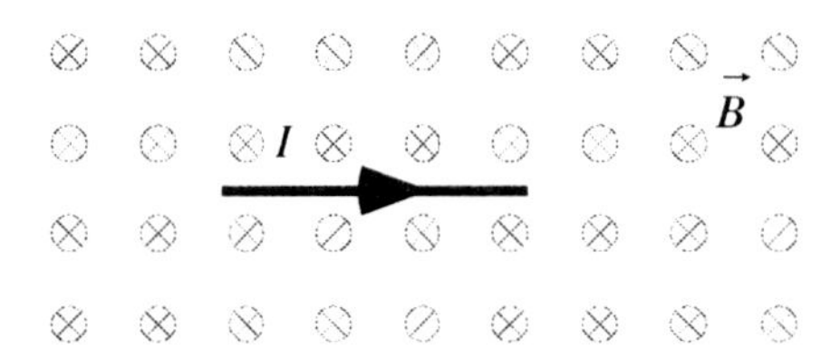

1) **What is the direction of the magnetic force acting on the wire segment due to the uniform magnetic field?**

2) **What would the direction of the magnetic force acting on the wire segment be if the direction of the uniform magnetic field were out of the paper?**

3) **What would happen to the magnitude of the magnetic force acting on the wire segment if the wire segment were longer but still completely within the uniform magnetic field?**

4) **What would happen to the direction of the magnetic force acting on the wire segment if we reversed the direction of current in the wire segment?**

5) **What would happen to the magnitude of the magnetic force acting on the wire segment if the wire segment were moved without changing its orientation so that its length was half-in and half-out of the uniform magnetic field region?**

MT4-QRT2: Current–Carrying Wire in a Uniform Magnetic Field—"Bend" of Wire

The figure below shows the effect of magnetic forces on the shape of a thin, flexible current-carrying wire with one end fixed, in a region of uniform magnetic field strength. The wire was initially straight.

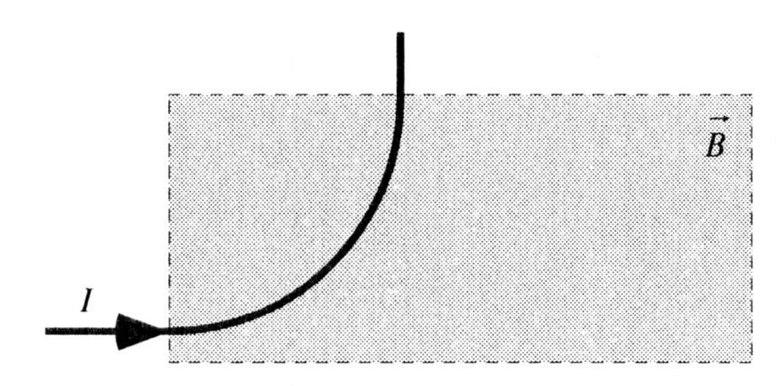

1) What is the direction of the external magnetic field acting on the wire?

2) If we reversed the direction of the external magnetic field, what would happen to the "bend" of the wire within the external field?

3) If we increase the current flowing in the wire, what will happen to the "bend" of the wire within the external magnetic field?

4) If we double the strength of the external magnetic field, what will happen to the "bend" of the wire within the external field?

5) If we reverse the direction of the current, what will happen to the "bend" of the wire within the external magnetic field?

MT5-QRT1: Two Parallel Long Wires—Force Difference

Two long wires are placed parallel to each other. The current in the top wire is I and the current in the bottom wire is $4I$.

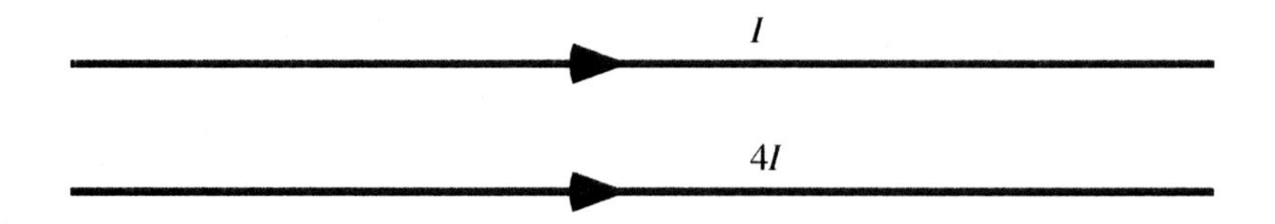

Described below are a number of changes to this situation. **For each change to this original situation, explain how the change will affect the force on the top wire, the force on the bottom wire, and the difference between the force the top wire exerts on the bottom wire and the force the bottom wire exerts on the top wire.**

	Force on top wire due to the bottom wire	Force on bottom wire due to the top wire	Difference between force the top wire exerts on the bottom wire and the force the bottom wire exerts on the top wire
The current in the top wire is doubled			
The current in the bottom wire is tripled			
The direction of the current in the bottom wire is reversed			
The wires are moved farther apart			

Carefully explain your reasoning.

MT5-QRT2: Suspended Permanent Magnet and Circular Coil—Scale Reading

A small permanent magnet is suspended by a spring balance above the center of a circular coil of wire that is sitting on a balance. A large current is now introduced into the coil causing the magnet to be attracted to the coil.

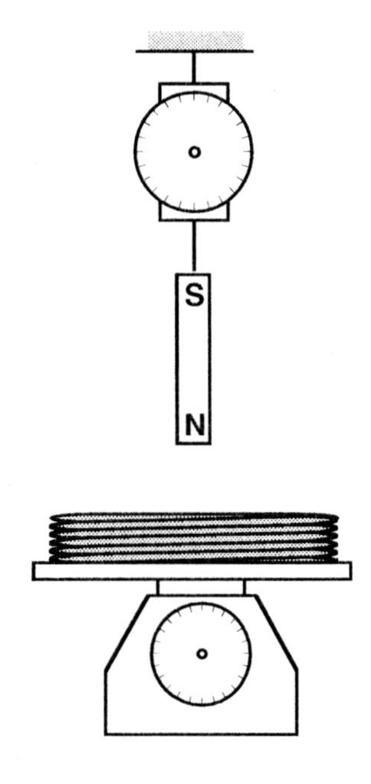

1) **Will the reading on the upper spring balance *increase, decrease,* or *stay the same*? Explain.**

2) **Will the reading on the balance supporting the coil *increase, decrease,* or *stay the same*? Explain.**

3) **Compare the sizes of the changes that will be observed in parts 1 and 2.**

MT6-QRT1: STRAIGHT CURRENT–CARRYING WIRE—MAGNETIC FIELD

The figure below shows a point *P* near a long current-carrying wire.

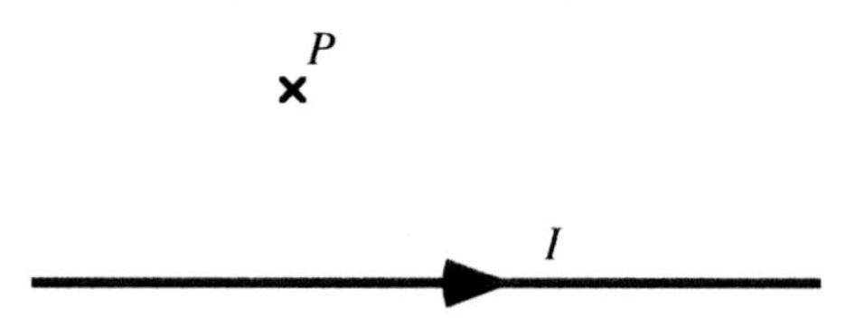

1) **What is the direction of the magnetic field at point *P* due to the current in the wire?**

2) **What would the direction of the magnetic field at point *P* be if the current in the wire were reversed?**

3) **What would happen to the magnetic field at point *P* if the current in the wire were increased?**

4) **What would happen to the magnetic field at *P* if point *P* were farther away from the wire?**

MT8-QRT1: THREE PARALLEL CURRENT–CARRYING WIRES I—MAGNETIC FIELD

All three wires shown have current *I*. The current in wire *Z* is into the page. The magnetic fields at the midway points between adjacent wires are indicated.

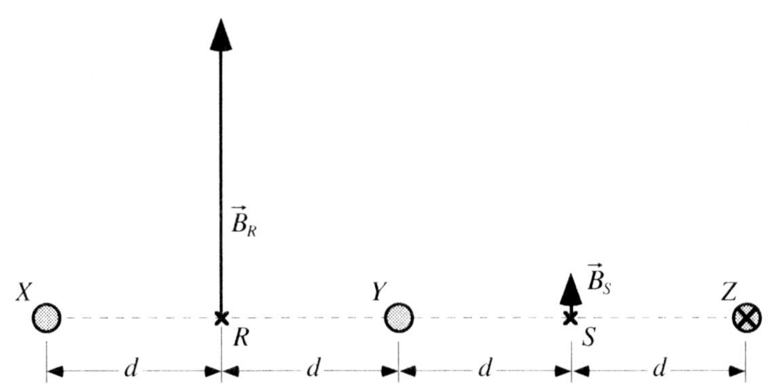

1) What is the direction of the current in wire *X*? Explain.

2) What is the direction of the current in wire *Y*? Explain.

For the situation shown above, describe how each of the following changes will affect the total magnetic field at the point midway between wires *X* and *Y*.

3) The current in wire Y is reversed. Explain.

4) The currents in wires X and Z are doubled. Explain.

5) The current in Y goes to zero. Explain.

6) The currents in all three wires are reversed. Explain.

MT8-QRT2: THREE PARALLEL CURRENT–CARRYING WIRES II—MAGNETIC FIELD AT A WIRE

Three long straight wires are parallel to each other and have the same current I into or out the page as indicated.

For the situation shown above, describe how each of the following changes will affect the total magnetic field acting on wire *Y*.

1) The current in wire *Y* is reversed.

2) The directions of the currents in *Y* and *Z* are both reversed.

3) The currents in wires *X* and *Z* are doubled.

4) The current in wire *Y* is reduced to one-third of original.

5) The direction of the current in *X* is reversed.

MT10-QRT1: MOVING CHARGE IN A UNIFORM MAGNETIC FIELD—KINETIC ENERGY CHANGE

A negatively charged particle enters a region in which there is a uniform magnetic field. The velocity of the particle is initially perpendicular to the field.

Explain how each of the changes described below would affect the particle's change of kinetic energy during the first 3 seconds it is in the field.

1) **The initial speed of the particle is three times as great.**

2) **The magnetic field direction is reversed.**

3) **The particle is replaced with a particle of identical charge, but double the mass.**

4) **The particle is initially moving at an angle of 60° relative to the direction of the magnetic field.**

MT11-QRT1: Moving Rectangular Loops in Uniform Magnetic Fields—Magnetic Flux and Flux Change

A rectangular wire loop is being pushed into a region of uniform magnetic field at a constant speed. At the instant shown, 20% of the loop is within the magnetic field.

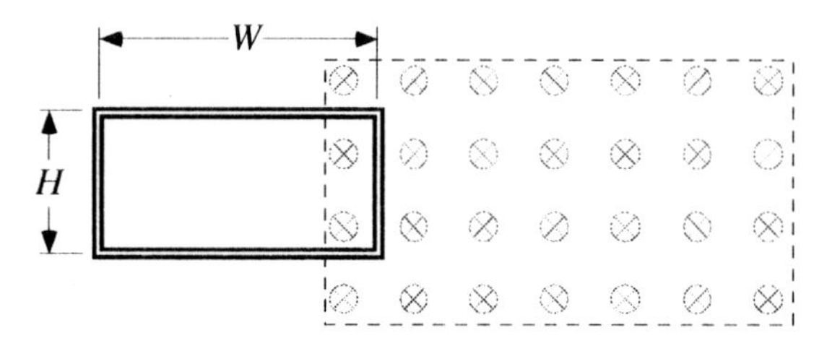

As compared to the situation shown above, indicate whether the following changes would *increase, decrease,* or *have no effect* on (1) the magnetic flux through the loop; and (2) the time rate of change of the magnetic flux through the loop.

	Change to the original situation	Magnetic flux	Time rate of magnetic flux change
1)	**The width, *W*, of the loop is increased so more of the loop is now in the field.**		
2)	**The height, *H*, of the loop is cut in half.**		
3)	**The height, *H*, of the loop is cut in half and the field strength is doubled.**		
4)	**The loop is completely inside the field.**		
5)	**The speed at which the loop is moving is greater.**		
6)	**The speed of the loop increases and the field strength increases.**		
7)	**The loop is leaving the field rather than entering it with 20% of the loop still within the field.**		
8)	**The strength of the magnetic field that the loop is moving into is greater.**		
9)	**The loop is moved farther to the right so that 40% of the loop is now within the field.**		

MT11-QRT2: Moving Rectangular Loops in Uniform Magnetic Fields—Magnetic Flux and Flux Change

A rectangular wire loop is being pushed into a region of uniform magnetic field at a constant speed.

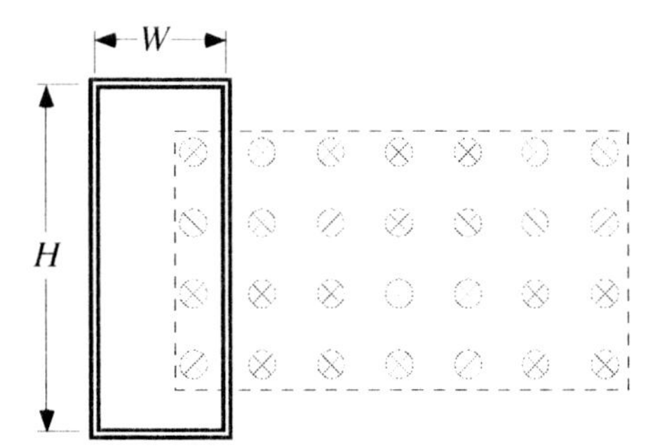

As compared to the situation shown above, indicate whether the following changes would *increase, decrease,* or *have no effect* on (1) the magnetic flux through the loop; and (2) the time rate of change of the magnetic flux through the loop.

	Change to the original situation	Magnetic flux	Time rate of magnetic flux change
1)	**The speed at which the loop is moving is increased.**		
2)	**The height, *H*, of the loop is increased.**		
3)	**The strength of the magnetic field is increased.**		
4)	**The loop has moved farther to the right so that more of the loop but not all is now within the field.**		
5)	**The width, *W*, of the loop is decreased so less of the loop is now in the field.**		
6)	**The loop has moved farther to the right so that both sides of the loop are within the field.**		
7)	**The height, *H*, of the loop is increased and the field strength is doubled.**		
8)	**The loop is leaving the field rather than entering it.**		
9)	**The speed of the loop is increased and the field strength is also increased.**		

MT13-QRT1: Changing Current—Bulb Brightness

Two circular loops of wire with small bulbs in them are sitting beside two long straight current-carrying wires. In both cases, these loops are the same distance away from the current-carrying wires. The wire loops, bulbs, and long straight wires are identical for the two situations but the currents in the straight wires are different. A graph of current in each wire as a function of time is shown at the right below.

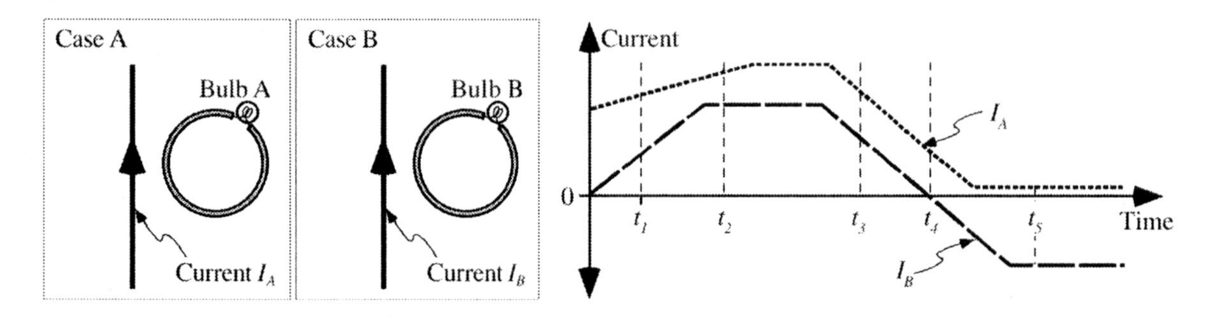

For each labeled time t_1-t_5, state whether the brightness of bulb A is *greater than, less than,* or *equal to* the brightness of bulb B. If either bulb is not lit at all at that time, state that explicitly.

1) **Time t_1:**

2) **Time t_2:**

3) **Time t_3:**

4) **Time t_4:**

5) **Time t_5:**

MT13-QRT2: Circular Loop outside a Long Solenoid—Induced Current

A wire loop is placed outside of and concentric with a long, tightly wound solenoid as shown. The solenoid has an initial current I. The current decreases to half its initial value in a time interval Δt. This change will produce an induced current I_{Ind} in the wire loop.

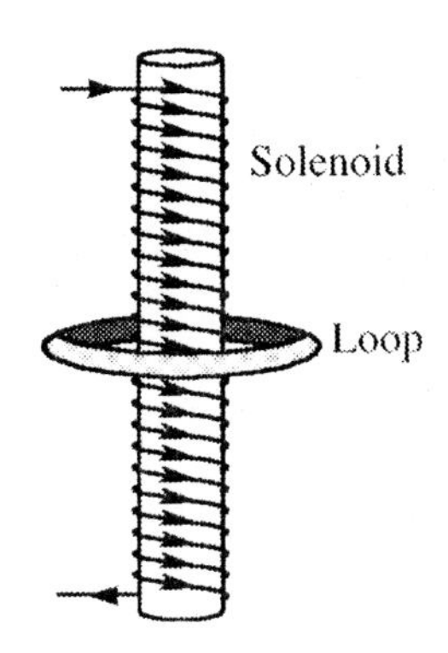

Explain how each of the changes described below would affect the induced current in the wire loop.

1) The decrease in the solenoid current occurs in a longer time.

2) The diameter of the wire loop is doubled.

3) The diameter of the solenoid is increased but the solenoid is still inside the wire loop.

4) The initial solenoid current is three times the value above but decreases to half its initial value in the same time interval.

MT14-QRT1: Loop Moving in a Uniform Magnetic Field—Induced and Total Magnetic Field

A rectangular wire loop is moving into a region of uniform external magnetic field at a constant speed. At the instant shown, 20% of the loop is within the magnetic field.

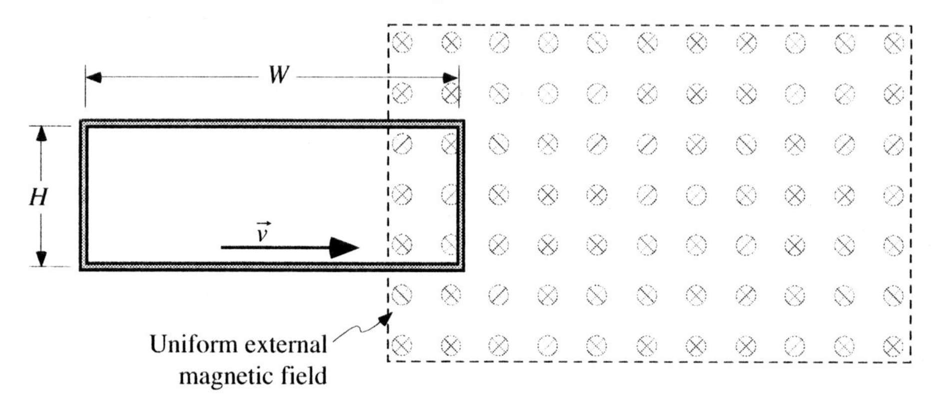

As compared to the original situation, would the following changes *increase*, *decrease*, or *have no effect* on magnitudes of the (1) total magnetic field and (2) the induced magnetic field at the center of the loop?

	Change to the original situation	Total B	Induced B
1)	**The speed at which the loop is moving is greater.**		
2)	**The strength of the magnetic field that the loop is moving into is greater.**		
3)	**The loop has moved farther to the right so that 40% of the loop is now within the field.**		
4)	**The loop has moved farther to the right so that 60% of the loop is now within the field.**		
5)	**The loop is moved farther to the right so that it is entirely within the field.**		
6)	**The loop is leaving the field rather than entering it with 80% of the loop still within the field.**		
7)	**The loop is leaving the field rather than entering it with 20% of the loop still within the field.**		

MT14-QRT2: Loops and Magnetic Field—Direction of Induced Magnetic Field

Identical rectangular wire loops are moving with velocities and accelerations as shown into, within, or out of a region in which there is a uniform external magnetic field coming out of the page. At the instant shown, various portions (25%, 75%, or 100%) of the loops are within the field.

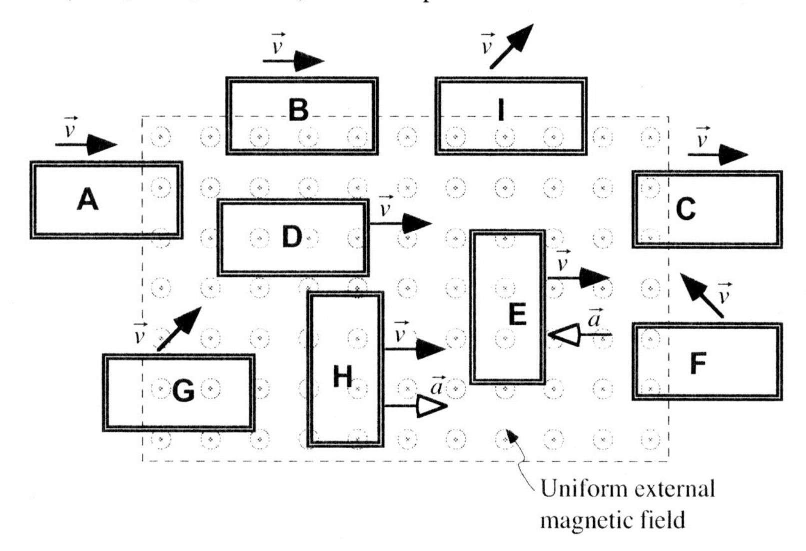

Select the direction of the induced magnetic field at the center of the rectangular loops from the choices 1-10 at the right. If there is no induced magnetic field, explicitly write "none" and if it cannot be determined from the information presented, write "unknown." Assume there is no interaction between the loops. Note that the acceleration is zero except for loops E and H.

Loop **Direction of the induced magnetic field at center of loop**

A ____________

B ____________

C ____________

D ____________

E ____________

F ____________

G ____________

H ____________

I ____________

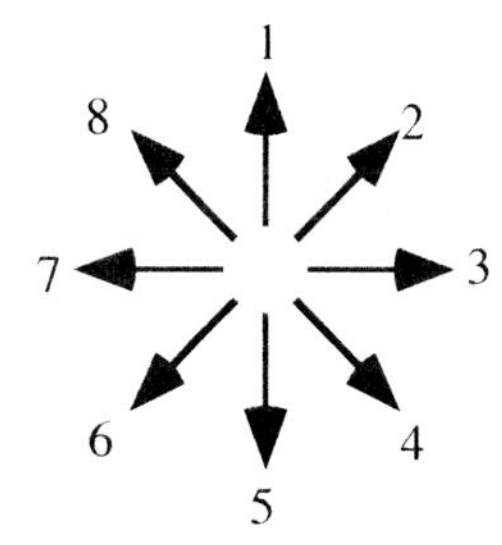

⊗ 9 (Into page)

⊙ 10 (Out of page)

MT15-QRT1: WIRE LOOPS AND MOVING MAGNETS—MOTION OF THE SYSTEM

A circular loop of wire is suspended from a thread so that it hangs freely. A permanent bar magnet is moved toward the center of the wire loop as shown.

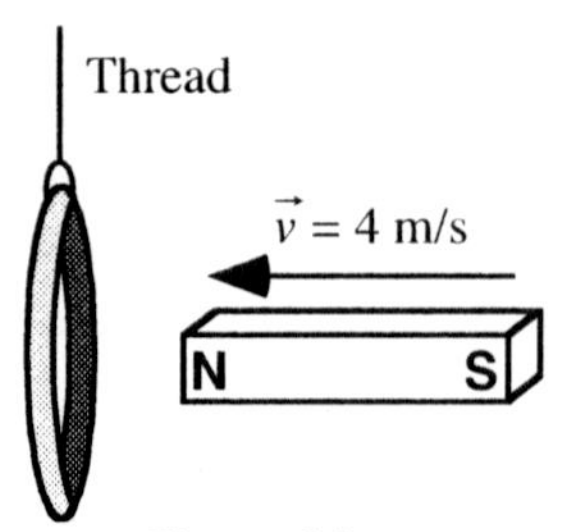

Describe how each of the following changes affects this system.

1) **The magnet is moving toward the loop at twice the speed.**

2) **A small gap is cut in the wire loop.**

3) **The south pole of the magnet is on the side of the magnet closer to the loop.**

4) **The north pole is moving toward the loop from the left.**

5) **The strength of the magnet is increased.**

6) **The magnet is moving away from the loop at the same speed.**

Linked Multiple Choice Tasks (LMCT)

MT2-LMCT1: Moving Charge within a Uniform Magnetic Field—Force

A charged particle moving at a constant speed is entering a region in which there is a uniform magnetic field. The particle follows the curved path shown.

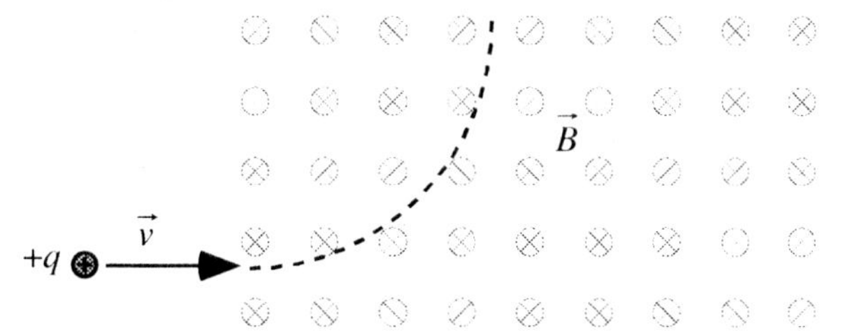

A number of changes to this initial situation are described in 1-6 below. Select from choices a-e how each change will affect the magnetic force on the particle shortly after it enters the magnetic field.

This change will:

(a) alter only the direction of the force on the particle.

(b) only increase the magnitude of the magnetic force on the particle.

(c) only decrease the magnitude of the magnetic force on the particle.

(d) alter both the magnitude and direction of the magnetic force on the particle.

(e) not affect the magnetic force on the particle.

(f) cause the magnetic force on the particle to be zero.

Each change below refers to the initial situation described above:

1) The $+q$ particle is replaced by a $+2q$ particle. ______

2) The $+q$ particle is replaced by a $-q$ particle. ______

3) The $+q$ particle is replaced by a neutral particle. ______

4) The particle enters the region moving at a slower initial velocity. ______

5) The magnetic field is one-third its original strength. ______

6) The direction of the magnetic field is parallel to the particle's initial velocity. ______

MT3-LMCT1: MOVING CHARGE BETWEEN TWO CURRENT–CARRYING WIRES—ACCELERATION

At the instant shown, a particle with a charge of +7 nC is located midway between two long parallel wires that have currents of 8 A and 4 A.

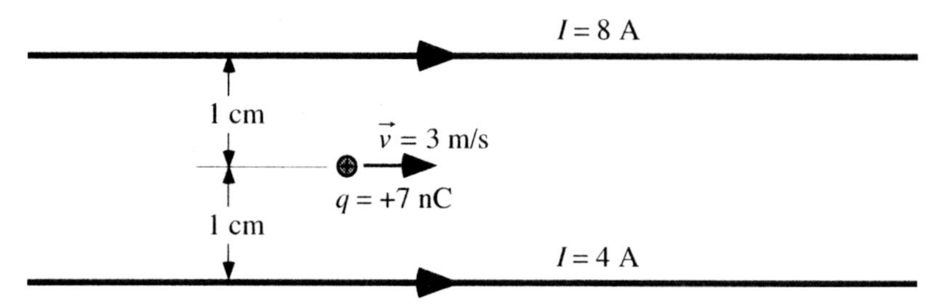

A number of changes to this initial situation are described in 1-8 below. Select from choices a-e how each change will affect the initial acceleration of the particle.

This change will:

(a) not affect the initial acceleration.

(b) increase the magnitude of the initial acceleration but not affect its direction.

(c) decrease the magnitude of the initial acceleration but not affect its direction.

(d) alter the direction of the initial acceleration but not affect its magnitude.

(e) alter both the magnitude and direction of the initial acceleration.

Each change below refers to the initial situation described above:

1) The current in both wires is doubled. _______

2) The direction of the current in the lower wire is reversed. _______

3) The +*q* particle is replaced with another particle having twice the charge and twice the mass. _______

4) The +*q* particle is replaced with a –*q* particle having a charge of –7 nC with the same mass. _______

5) The +*q* particle is replaced with another particle having twice the mass with the same charge. _______

6) The initial velocity of the particle is doubled. _______

7) The wires are both moved farther away from the charged particle. _______

8) The current in the lower wire is increased to 10 A. _______

mT3-LMCT2: Charge Moving Along Wire—Magnetic Force

At the instant shown, a particle with a charge of $+q$ is a distance d from a long straight wire and is moving parallel to the wire.

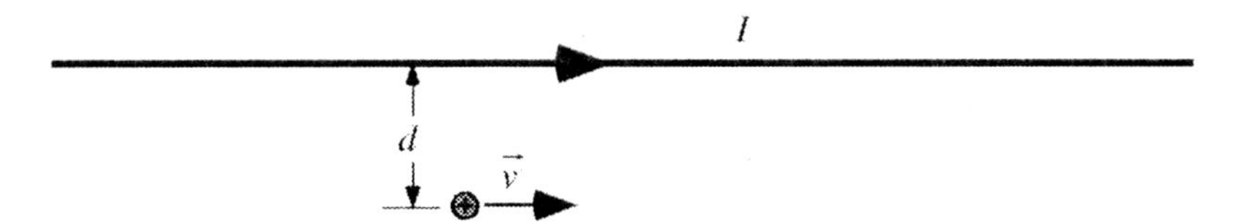

A number of changes to this initial situation are described in 1-6 below. Select from choices a-e how each change will affect the magnetic force on the particle at this instant.

This change will:

(a) have **no effect** on the force.

(b) **increase** the strength (magnitude), but not affect the direction of the force.

(c) **decrease** the strength, but not affect the direction of the force.

(d) **alter the direction**, but not affect the strength of the force.

(e) **alter both** the strength and direction of the force on the particle.

Each change below refers to the initial situation described above:

1) The current in the wire is doubled. ______

2) The direction of the current in the wire is reversed. ______

3) The $+q$ particle is replaced with a $+2q$ particle with the same mass. ______

4) The $+q$ particle is replaced with a $-q$ particle with the same mass. ______

5) The $+q$ particle is replaced with a particle having triple the mass and the same charge. ______

6) The charged particle's initial velocity is toward the bottom of the page. ______

MT4-LMCT1: Current in a Uniform Magnetic Field—Magnetic Force

A section of straight wire within a magnetic field is conducting a current to the right. The external magnetic field is uniform and directed into the paper.

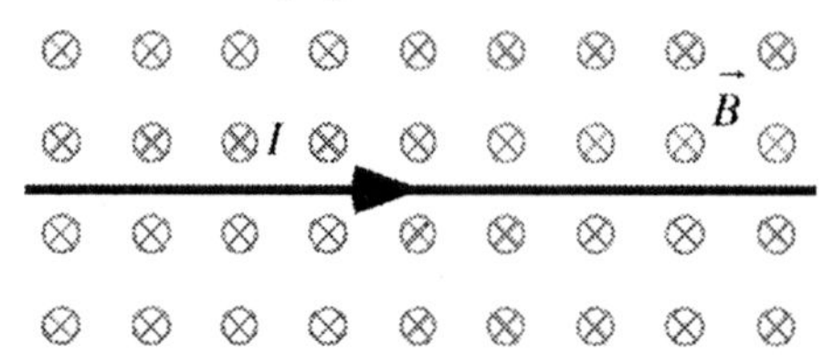

A number of changes to the initial force are described in 1-5 below. Select from choices a-g the possible cause of the change in the force.

This change could be caused by:

(a) **increasing** the **current**.

(b) **decreasing** (but not to zero) the **current**.

(c) **reversing** the direction of **current**.

(d) **increasing** the strength of the **magnetic field**.

(e) **decreasing** (but not to zero) the strength of the **magnetic field**.

(f) **reversing the direction** of the **magnetic field.**

(g) **none of these**.

If more than one choice is correct, please indicate ALL correct choices for the answer.

Each change below refers to the initial situation described above:

1) The magnetic force on the wire is larger and in the same direction. _______

2) The magnetic force on the wire is larger and in the opposite direction. _______

3) The magnetic force on the wire is smaller and in the same direction. _______

4) The magnetic force magnitude remains the same, but the direction changes. _______

5) The magnetic force on the wire is zero. _______

MT6-LMCT1: LONG WIRE WITH A CURRENT—MAGNETIC FIELD

Point P is located above a long straight wire that has a current to the right.

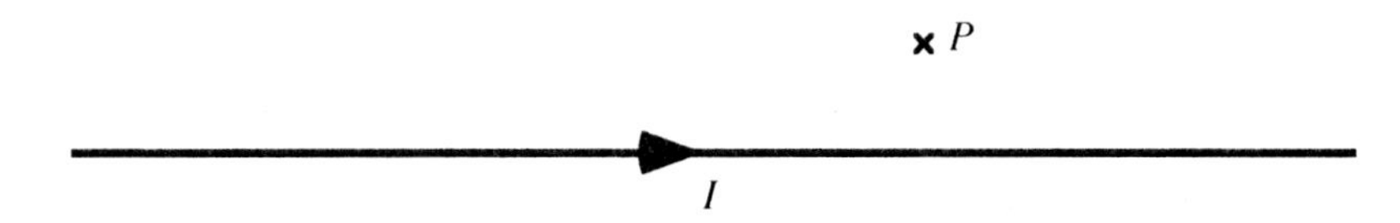

A number of changes to the initial magnetic field at point P are described in 1-5 below. Select from choices a-g what might have caused this change in the magnetic field at point P.

This change could be caused by:

(a) an **increase** in the current.

(b) a **decrease** in the current.

(c) **reversing** the direction of the current.

(d) moving P **farther away** from the wire toward the top of the page.

(e) moving P **closer** to the wire but still on the same side of the wire.

(f) moving P **below the wire** and the **same distance** from the wire.

(g) **none of these**.

If more than one choice is correct, please indicate ALL correct choices for the answer.

Each change below refers to the initial situation described above:

1) **The magnetic field at P decreases, but the direction remains the same.** _______

2) **The direction of the magnetic field at P is reversed.** _______

3) **The magnetic field at P is decreased and the direction is reversed.** _______

4) **The magnetic field at P points toward the right.** _______

5) **The magnetic field at P increases and points toward the top of the page.** _______

MT7-LMCT1: CURRENT–CARRYING CIRCULAR LOOP—MAGNETIC FIELD

A circular loop is conducting current in a counter-clockwise direction as shown.

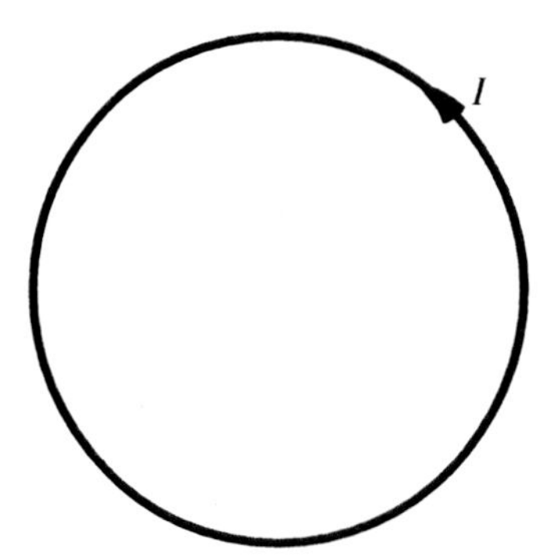

A number of changes to this initial situation are described in 1-6 below. Select from choices a-e how each change will affect the magnetic field at the center of the loop.

This change will:

(a) alter the direction of the magnetic field at the center of the loop.
(b) increase the magnitude of the magnetic field at the center of the loop.
(c) decrease the magnitude of the magnetic field at the center of the loop.
(d) alter both the magnitude and direction of the magnetic field at the center of the loop.
(e) not affect the magnetic field at the center of the loop.

Each change below refers to the initial situation described above:

1) The current in the loop is increased. _______

2) The direction of the current in the loop is reversed. _______

3) The current in the loop is reduced and its direction is reversed. _______

4) The radius of the loop is increased. _______

5) The radius of the loop is decreased and the current is increased. _______

6) The loop is replaced by a coil with several turns with the same radius and current. _______

MT8-LMCT1: THREE CURRENT–CARRYING WIRES—MAGNETIC FIELD BETWEEN WIRES

Three straight long wires have currents coming out of the page. Also shown is a point P between two wires.

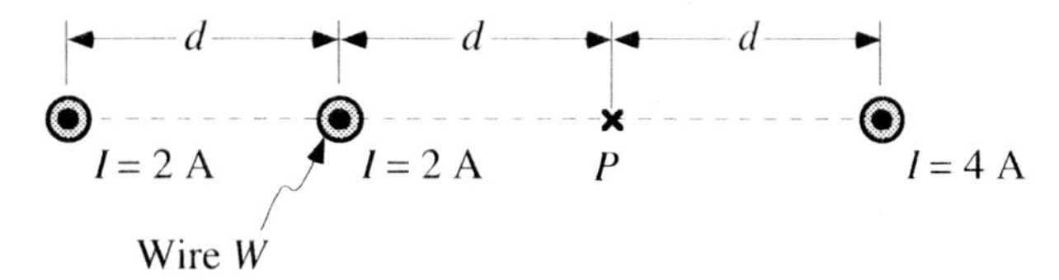

A number of changes to this initial situation are described in 1-7 below. Select from choices a-e how each change will affect the magnetic field produced at *P* by only the wire labeled *W*.

This change will:

(a) alter the direction of the field at P produced by the current in W.

(b) increase the magnitude of the field at P produced by the current in W.

(c) decrease the magnitude of the field at P produced by the current in W.

(d) alter both the direction and the magnitude of the field at P produced by the current in W.

(e) not affect the field at P produced by the current in W.

Each change below refers to the initial situation described above:

1) The current in the leftmost wire is doubled. _______

2) The rightmost wire is moved closer to point *P*. _______

3) The current in wire *W* is doubled. _______

4) The current in the right hand wire is reversed. _______

5) Wire *W* and the leftmost wire are interchanged. _______

6) Wire *W* is moved so that it is directly above *P* a distance *d* away. _______

7) The leftmost and rightmost wires are both moved a distance *d*/2 to the left. _______

MT8-LMCT2: THREE PARALLEL CURRENT–CARRYING WIRES I—MAGNETIC FIELD

The arrows in the figure represent the total magnetic field at points midway between wires X and Y and wires Y and Z. All three wires have the same current directed into the page.

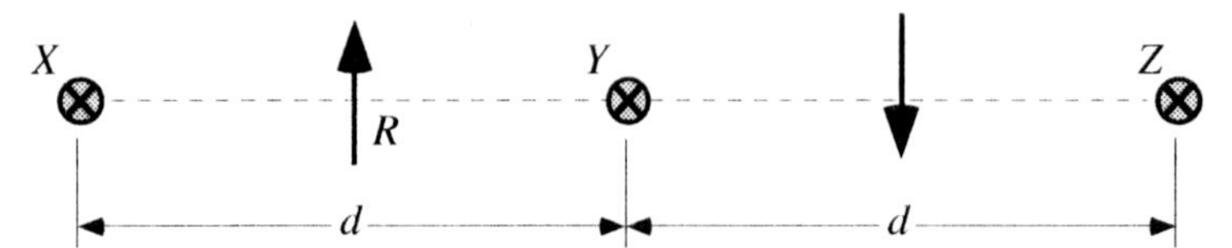

A number of changes to this initial situation are described in 1-6 below. Select from choices a-e how each change will affect the magnetic field midway between X and Y.

This change will:

(a) increase the magnitude of the field without changing its direction.

(b) decrease the magnitude of the field without changing its direction.

(c) reverse the direction of the field without changing its magnitude.

(d) change both the magnitude and direction of the field.

(e) have no effect on the magnetic field.

Each change below refers to the initial situation described above:

1) The current in wire Z is reversed. _______

2) The currents in all three wires are doubled. _______

3) The currents in both X and Y are reversed. _______

4) The current in wire Y is reduced to zero. _______

5) The current in wire Y is cut by 10%. _______

6) The currents in wires X and Z are reversed and cut in half. _______

MT9-LMCT1: THREE PARALLEL CURRENT–CARRYING WIRES I—MAGNETIC FORCE ON WIRE

Wires *A* and *B* have 4 A currents coming out of the page and wire *C* has a 4 A current going into the page. All the wires are parallel and equally spaced.

⊙ A - - - - - - - - ⊙ B - - - - - - - - ⊗ C

A number of changes to this initial situation are described in 1-6 below. Select from choices a-e how each change will affect the magnetic force on the center wire *B*.

This change will:

(a) increase the magnitude of the force without changing its direction.

(b) decrease the magnitude of the force without changing its direction.

(c) reverse the direction of the force without changing its magnitude.

(d) change both the magnitude and direction of the force.

(e) have no effect on the magnetic force.

Each change below refers to the initial situation described above:

1) The current in wire *B* is reversed. _______

2) The currents in all three wires are doubled. _______

3) The currents in the wires *A* and *C* are both reversed. _______

4) The current in wire *B* is reduced. _______

5) The currents in the wires *A* and *C* are reversed and cut in half. _______

6) Both *A* and *C* are moved the same distance so that they are closer to *B*. _______

MT12-LMCT1: MOVING RECTANGULAR LOOPS IN UNIFORM MAGNETIC FIELDS—EMF

A rectangular wire loop with a very small resistor (not shown) has its leading edge just moving into a uniform magnetic field. The loop has dimensions of H by W, and is moving at a constant speed.

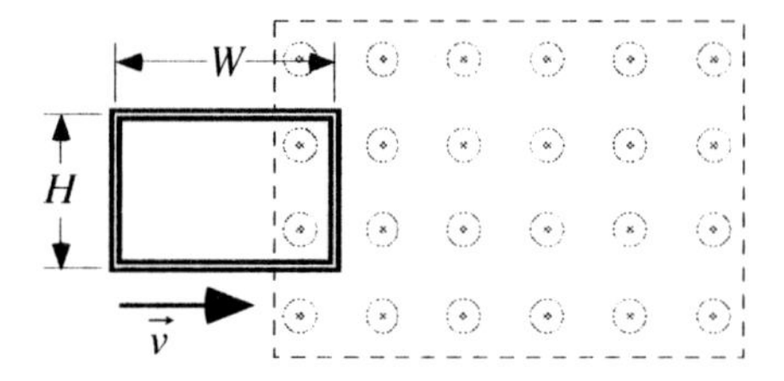

Described below are a variety of changes to this initial situation. All of the changes described are modifications of this initial situation.

Select from choices (a)-(f) how each change will affect the emf across the resistor in the loop.

The emf will:

(a) increase.

(b) decrease.

(c) reverse polarity.

(d) increase and reverse polarity.

(e) decrease and reverse polarity.

(f) not change.

Changes in the situation are:

1) **The magnetic field is doubled in strength.** __________

2) **The velocity of the loop is reduced.** __________

3) **The width, W, of the loop is tripled.** __________

4) **The direction of the magnetic field is reversed.** __________

5) **The height, H, of the loop is halved.** __________

6) **The direction of the magnetic field is reversed and the magnitude of the magnetic field is reduced.** __________

MT12-LMCT2: Rectangular Loop in a Uniform Magnetic Field—Velocity

A rectangular wire loop is pulled at a constant speed into a region in which there is a uniform magnetic field into the page.

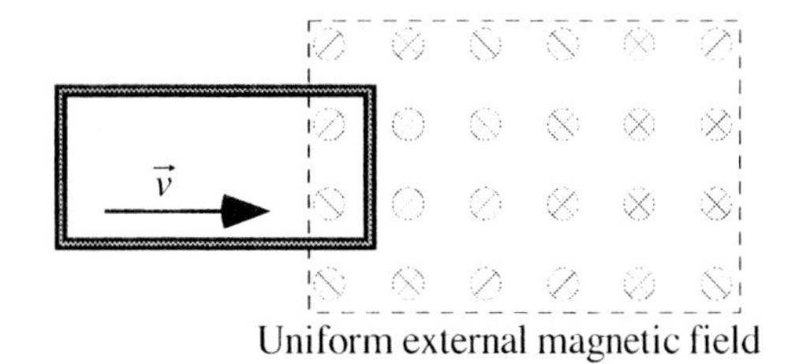

Uniform external magnetic field

Identify whether effects 1-7 listed below could result because the loop is moving at:

(a) a constant velocity approaching the region of the magnetic field but has not yet entered it.

(b) a constant velocity entering the region with 1/3 of the loop inside the region of the magnetic field.

(c) a constant velocity with the entire loop within the region of the magnetic field.

(d) a constant velocity leaving the region with 1/3 of the loop inside the region of the magnetic field.

(e) an increasing velocity with the entire loop within the region of the magnetic field.

If more than one choice is correct, please indicate all correct choices for the answer.

1) There is no magnetic flux in the loop. _______

2) The magnetic flux through the loop is increasing. _______

3) There is an induced emf producing a current in the loop. _______

4) There is an induced clockwise current in the loop. _______

5) There is an induced magnetic field within the loop. _______

6) There is an induced counter-clockwise current in the loop. _______

7) The magnetic flux through the loop is decreasing. _______

MT13-LMCT1: MOVING RECTANGULAR LOOPS IN UNIFORM MAGNETIC FIELDS—CURRENT

A rectangular wire loop containing a one-ohm resistor (not shown) is being pulled at a constant speed into a region in which there is a uniform magnetic field pointing into the page as shown in the figure below.

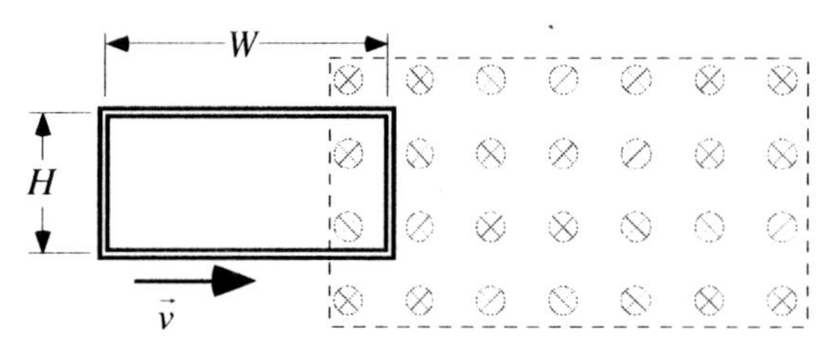

Indicate how the current in the loop will change if each of the following modifications are made to the original situation shown above.

Possible changes to the current are:

(a) The current will be larger.

(b) The current will be smaller but non-zero.

(c) The current will be the same magnitude.

(d) The current will be zero.

Changes in the situation are:

1) The loop is almost completely in the field. ______

2) The speed of the loop is tripled. ______

3) The field direction is reversed. ______

4) The height, *H*, of the loop is decreased. ______

5) The width, *W*, of the loop is increased so that more of the loop is within the field. ______

6) The loop is leaving the field region. ______

7) The field strength is decreased. ______

8) The loop's velocity at the instant shown has the same magnitude but points toward the bottom of the page. ______

MT13-LMCT2: LOOPS WITH BULBS NEAR A CURRENT—BULB LIGHTING

Two circular loops of wire with small bulbs in them are sitting beside two long straight current-carrying wires. The wire loops, bulbs, and long straight wires are identical for the two cases. The loops are the same distance from the straight wires.

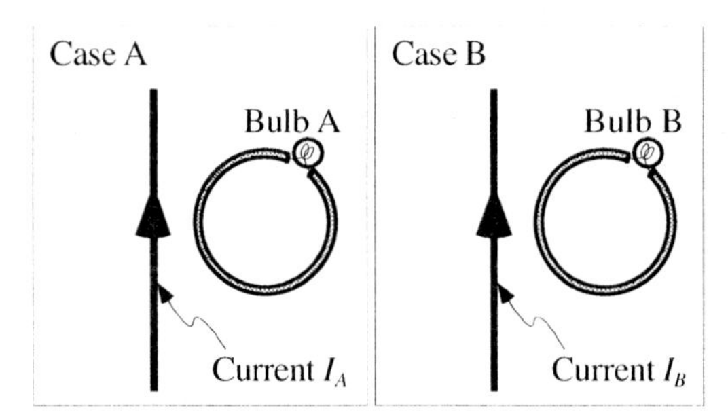

The currents in the long straight wires as a function of time are given in statements 1-6 below. For each of these situations, identify whether:

(a) neither bulb is lit.

(b) only bulb A is lit.

(c) only bulb B is lit.

(d) both bulbs are lit and equally bright.

(e) both bulbs are lit, and bulb A is brighter.

(f) both bulbs are lit, and bulb B is brighter.

For each situation 1-6 below identify the effect from a-f above.

1) **There is a constant current in wire A of 6 A (amperes) and in wire B of 15 A.** _______

2) **Both wires start with the same initial 3 A current, but the current in wire B increases to 6 A in a 0.3 second interval while that in A remains constant.** _______

3) **The current in wire A goes from 2 A to 10 A in a 0.5 second interval, while at the same time, the current in wire B goes from 12 A to 16 A.** _______

4) **The current in wire A decreases from 10 A to 4 A in a 0.2 second interval, while the current in wire B increases from 9 A to 18 A in 0.3 seconds.** _______

5) **The current in wires A and B both double in a 0.2 second interval, but the current in wire A starts at twice the initial value of B.** _______

6) **Both currents decrease from 32 A to 9 A, but the time interval for A is 0.5 seconds, while for B, the interval is 0.2 seconds.** _______

MT15-LMCT1: WIRE LOOPS AND MOVING MAGNETS—LOOP BEHAVIOR

A circular loop of wire is suspended by a thread. A permanent magnet is moved at a constant speed toward the center of the loop as shown. The loop of wire swings away from the magnet.

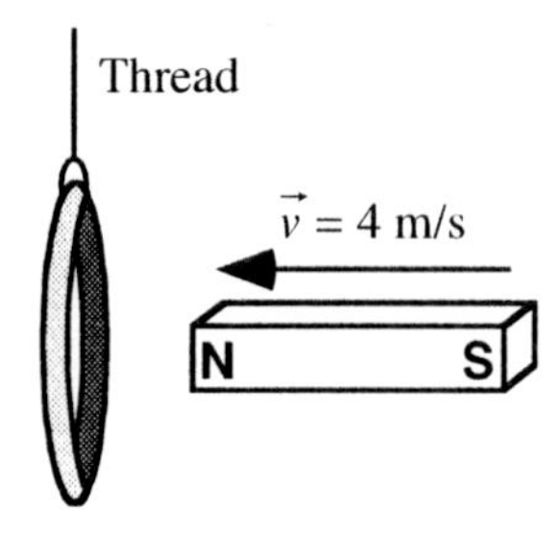

Listed below are a number of changes to this original situation.

The possible choices for how the change will affect the loop are:

(a) The loop will swing away from the magnet faster than the original case.
(b) The loop will swing away from the magnet slower than the original case.
(c) The loop will swing toward the magnet.
(d) There will be no change in the behavior of the loop.
(e) The loop will not move in this case.

For each change 1-6 below, identify how the behavior of the loop is altered.

1) The magnet is moved toward the loop with twice the velocity. ______

2) A small gap is cut in the loop. ______

3) The magnet is flipped around so the south pole is closer to the loop. ______

4) A plastic loop is used instead of a metal one. ______

5) A stronger magnet is used. ______

6) The magnet (with the north pole still near the loop) is moved away from the loop at the same speed. ______

CONFLICTING CONTENTIONS TASKS (CCT)

MT1-CCT1: ELECTRIC CHARGE NEAR A BAR MAGNET—FORCE DIRECTION

Consider the following students' statements about the magnetic force on a positively charged particle placed at rest near a permanent magnet.

Andrea: *"A positively charged particle placed near the north pole of a permanent magnet will experience a repulsive force because the north pole acts like a positive charge."*

Ben: *"A positively charged particle placed near the north pole of a permanent magnet will experience an attractive force, but not because it is a magnet."*

Chico: *"A positively charged particle placed near the north pole of a permanent magnet will not experience any electromagnetic force since the charge is not moving."*

Which of these students is correct?

Andrea _____ Ben _____ Chico _____ None of them _____ **Explain.**

MT1-CCT2: CHARGE NEAR A CIRCULAR CURRENT LOOP—MAGNETIC FORCE DIRECTION

Consider two cases where a positively charged particle is placed at rest near a circular wire loop. In case *A* there is no current in the loop, and in case *B* there is counter-clockwise current as viewed from the position of the charge. Consider the following students' statements:

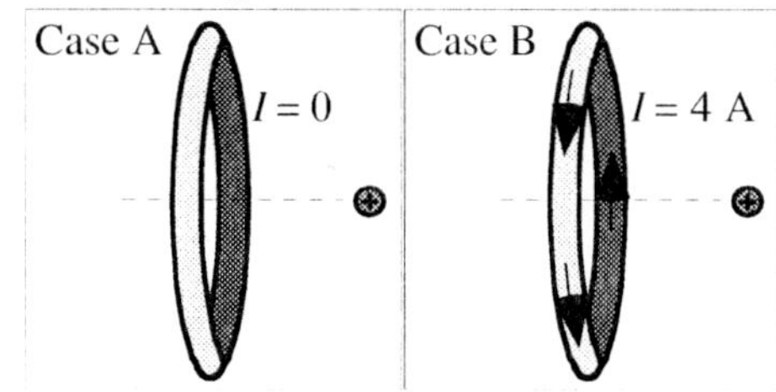

Arturo: *"There is no magnetic force in either case since neither charge is moving."*

Bonita: *"There will be no magnetic force in case A because there is no current. The particle will be attracted in case B due to the magnetic field produced by the current in the loop."*

Caleb: *"There is going to be a repulsion in case B since the current loop acts like a magnet. There is no magnetic force in case A since there is no current."*

Which of these students is correct?

Arturo _____ Bonita _____ Caleb _____ None of them _____ **Explain.**

MT3-CCT1: CHARGED PARTICLE AND STRAIGHT CURRENT–CARRYING WIRE—FORCE

Three students discuss the force on a charged particle moving near a long straight wire that is carrying a current.

Aaron: *"If the velocity and current are parallel, then the force is zero."*

Brenda: *"There is no force if the particle is moving straight away from the wire because the velocity is parallel to the magnetic field."*

Craig: *"As long as the particle is moving near the wire, the particle will experience a force."*

Which of these students is correct?

Aaron _____ Brenda _____ Craig _____ None of them _____ **Explain.**

MT5-CCT1: MOVING MAGNET AND CIRCULAR LOOP—FORCE

Three students are comparing the forces on a permanent magnet and a current-carrying circular loop of wire. The mass of the magnet is much larger than the mass of the loop.

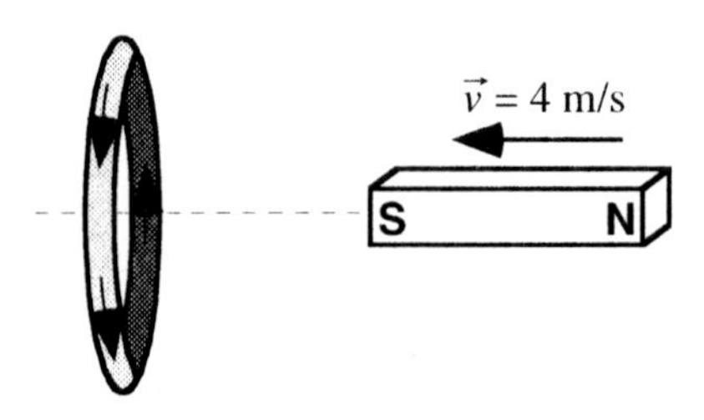

Anders: *"The coil will push or pull on the magnet just as hard as the magnet pushes or pulls on the coil."*

Bao: *"I think the magnet has to push harder on the coil than the coil pushes on the magnet because the magnet is more massive than the wire."*

Cho: *"I think the magnet will push or pull on the coil but the coil will not push or pull on the magnet at all because the coil is not a magnet."*

Which of these students is correct?

Anders_____ Bao_____ Cho_____ None of them_____ **Explain.**

MT5-CCT2: TWO MAGNETS—FORCE

Three students are discussing the strengths (or magnitudes) of the forces between two permanent magnets. The smaller magnet is moving to the right. The larger magnet is stronger than the smaller magnet.

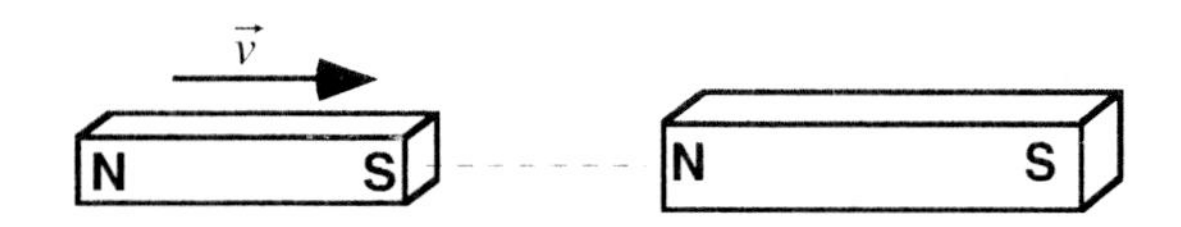

Alejandro: *"The velocities and magnet strengths don't matter. The magnets will attract each other with equal strength."*

Bernardo: *"No, the stronger magnet will push more than the weaker one."*

Cecilia: *"I don't think we can compare the strength of the forces unless we know the velocity of the smaller magnet."*

Which of these students is correct?

Alejandro_____ Bernardo_____ Cecilia_____ None of them____ **Explain.**

MT8-CCT1: THREE PARALLEL CURRENT–CARRYING WIRES II—FORCE

Three long parallel wires all have the same magnitude current. There is no force on wire Y. The distance between adjacent wires is the same. Three students are discussing this situation.

X Y Z

$I = 2$ A $I = 2$ A $I = 2$ A

Anita: *"For the force on wire Y to cancel, the currents in X and Z must be in opposite directions."*

Beatrice: *"No, for the forces on wire Y to cancel, the current in X and Z must be in the same direction, but opposite the current direction in Y."*

Carol: *"No, for the forces on wire Y to cancel, all we care about is the current in X and Z being in the same direction."*

Which of these students is correct?

Anita _____ Beatrice _____ Carol _____ None of them _____ **Explain.**

MT11-CCT1: Moving Rectangular Loops in Uniform Magnetic Fields—Magnetic Flux Change

Three students are discussing a rectangular wire loop moving at a constant speed into a region in which there is a uniform magnetic field. The sides of the rectangular wire loop are perpendicular or parallel to the leading edge of the magnetic field.

Alicia: *"If a rectangular wire loop is pulled into a region where there is a uniform magnetic field, the rate of change of magnetic flux through the wire loop will increase as more of the loop gets into the field."*

Banji: *"The rate of change of magnetic flux through the wire loop will be constant as the loop moves into the field region since the loop is moving at a constant speed."*

Colin: *"No, the rate of change of magnetic flux through the wire loop will be zero since the loop is moving at a constant speed."*

Which of these students is correct?
Alicia _____ Banji _____ Colin _____ None of these students _____ **Explain.**

MT12-CCT1: Moving Rectangular Loops in Uniform Magnetic Fields—Emf

Consider the following statements about the open rectangular loop (note the gap in wire) that is moving at a constant speed into a region with a uniform magnetic field.

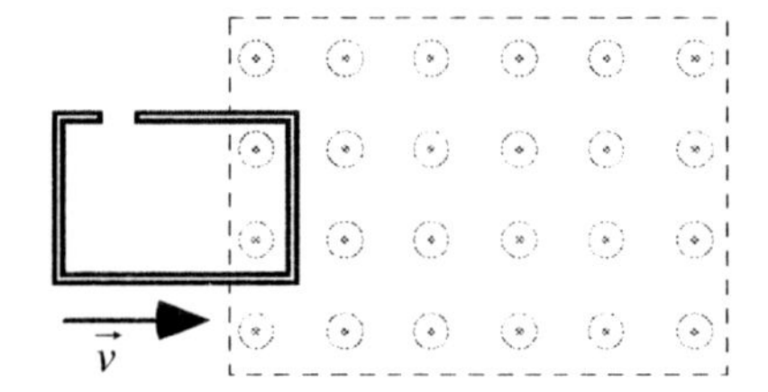

Three students contend:

Anna: *"An open rectangular loop will have an emf across the gap when moving into or out of a uniform magnetic field because that is when the magnetic flux is changing."*

Beth: *"An open rectangular loop will have an emf across the gap only when the gap itself is inside the magnetic field region."*

Carmen: *"An open rectangular loop cannot have an emf across the gap because there is no current in the loop."*

Which of these students is correct?
Anna _____ Beth _____ Carmen _____ None of these students _____ **Explain.**

MT13-CCT1: MOVING RECTANGULAR LOOPS IN UNIFORM MAGNETIC FIELDS—CURRENT

Three students are discussing a rectangular wire loop moving at a constant speed as it enters a region in which there is a uniform magnetic field perpendicular to the plane of the loop. The sides of the rectangular wire loop are perpendicular or parallel to the leading edge of the magnetic field.

Ahmed: *"If a rectangular wire loop is pulled into a region where there is a uniform magnetic field, the current will increase as more of the loop gets into the field since there will be more magnetic flux inside the loop."*

Bo: *"No, the current in the wire loop will decrease as the loop moves into the field region since less of the loop will be outside of the field."*

Cathy: *"No, the current in the wire loop will be constant from the time the loop starts into the field region until it is fully into the field region. Then the current will go to zero."*

Which of these students is correct?

Ahmed _____ Bo _____ Cathy _____ None of these students _____ **Explain**.

MT13-CCT2: CHANGING CURRENT—BULB BRIGHTNESS

In each case below, a circular loop of wire with a small bulb in it is placed beside a long straight current-carrying wire. In both cases, these loops are the same distance away from the current-carrying wire. Bulb A is brighter than bulb B. The wire loops, bulbs, and long straight wires are identical for the two situations.

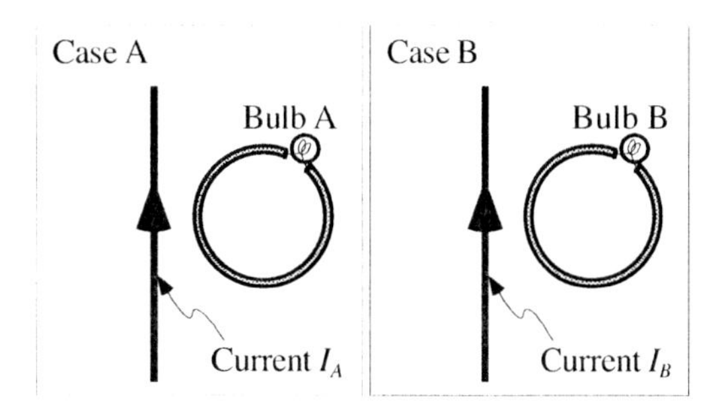

Three students contend:

Angel: *"Bulb A is brighter than bulb B because the long wire next to the brighter bulb has a larger current in it."*

Blanca: *"No, bulb A is brighter than bulb B because the current in the long wire next to it is changing at a faster rate than the current in the other wire."*

Caitlin: *"Bulb A is brighter because the current in the long wire next to it is increasing while the current in the long wire next to bulb B is decreasing."*

Which of these students is correct?

Angel _____ Blanca Caitlin None of these students **Explain.**

MT14-CCT1: Moving Loops in Uniform Magnetic Fields—Magnetic Field

A vertical conducting bar forms a complete rectangular loop with the conductor it is sliding on. The vertical bar is moving to the right at a constant speed in a region with a uniform external magnetic field. Points A and B are the same distance from the sliding bar at the instant shown.

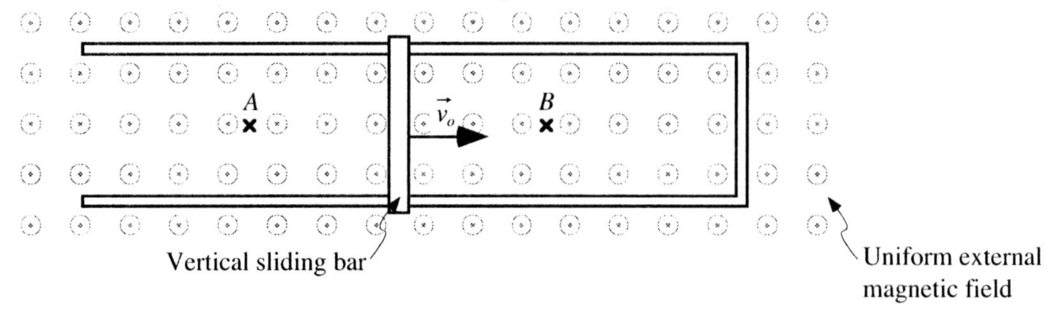

Three students make the following statements about the total magnetic field at A and B:

Antonio: *"I think the magnetic field at points A and B will be the same since they are the same distance from the vertical bar."*

Brooke: *"I think the magnetic field at point A is the same as the external field because it is outside the loop and at point B it will be larger."*

Carlos: *"No, I think the magnetic fields at points A and B will both be larger than the external field due to the induced magnetic field."*

Which of these students is correct?

Antonio ______ Brooke ______ Carlos ______ None of these students ______ **Explain.**

MT14-CCT2: LOOP MOVING INTO A UNIFORM MAGNETIC FIELD—INDUCED MAGNETIC FIELD

A rectangular wire loop is moving to the right at a constant velocity into a region in which there is a uniform external magnetic field coming out of the page. Labels *A* and *B* indicate points that are the same distance from the loop.

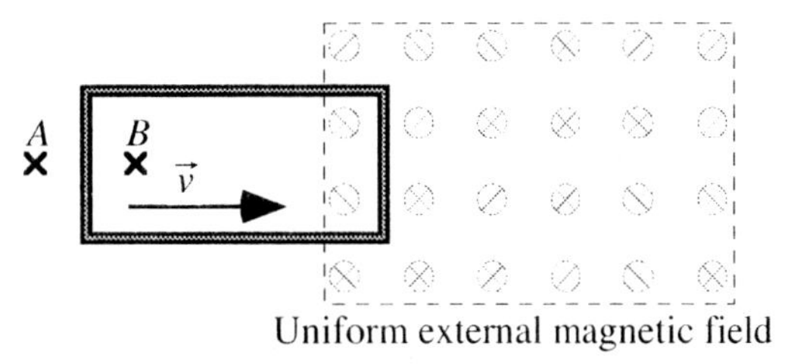

Uniform external magnetic field

Three students contend at the instant the loop is at the position shown:

Amber: *"The magnetic field at points A and B will be the same but in opposite directions since they are the same distance from the wire loop."*

Brian: *"I think the magnetic field at point B is larger than at point A since it is inside the loop. The fields at A and B will also be in opposite directions."*

Christina: *"No, I think the magnetic field at points A and B will both be zero since they are outside the region with the uniform magnetic field."*

Which of these students is correct?

Amber _____ Brian _____ Christina _____ None of these students _____ **Explain.**

CHANGING REPRESENTATIONS TASKS (CRT)

MT2-CRT1: CHARGE IN A UNIFORM MAGNETIC FIELD EQUATION—ACCELERATION GRAPH

The equation below represents at an instant the magnetic force on a charged particle moving in a uniform magnetic field.

$$-6.49\times10^{-11}\hat{\mathbf{j}}\text{N} = (1.602\times10^{-19}\text{C})(2.50\times10^{7}\hat{\mathbf{i}}\text{m/s})\times(0.162\hat{\mathbf{k}}\text{T})$$

Draw and label an appropriate diagram showing the magnetic field, the particle's velocity, and the force on the particle at this instant.

Sketch the shape of the graph of the magnitude of the acceleration of the charged particle versus time for the time that the particle is in the magnetic field. It is not required to put values on the vertical axis.

MT3-CRT1: Long Current–Carrying Wire II—Magnetic Field

The four labeled points in the figure below are in the vicinity of a long wire with current *I*. The magnetic field at point *A* has a magnitude of 24 μT.

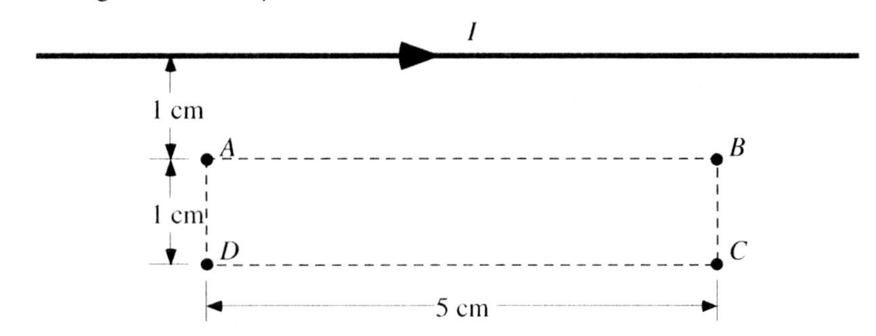

Graph the magnitude of the magnetic field along the dashed line paths from *A* to *B* to *C* to *D* and then back to *A*.

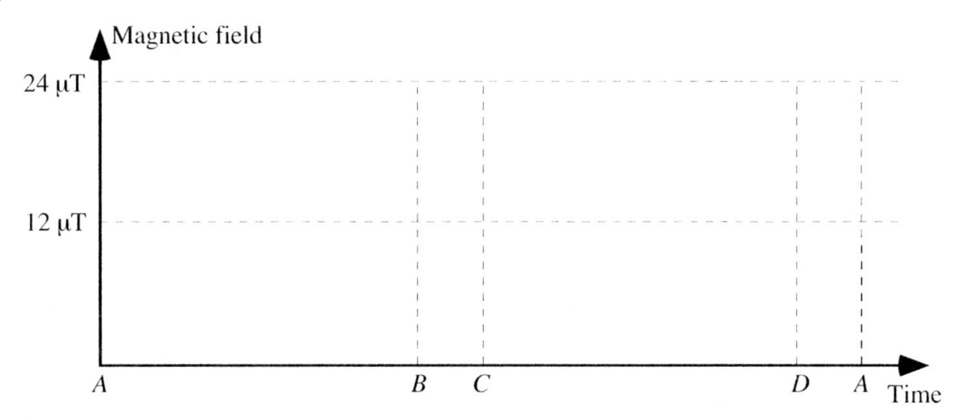

MT4-CRT1: Force Equation—Diagram of the Current in a Magnetic Field

The equation below represents the magnetic force on a current-carrying straight wire segment in a uniform magnetic field.

$$-0.648\hat{\mathbf{j}}\ \mathrm{N} = (1.60\hat{\mathbf{i}}\ \mathrm{A})(2.50\ \mathrm{m})\ \times\ (0.162\hat{\mathbf{k}}\ \mathrm{T})$$

Draw and label an appropriate diagram of the current-carrying wire and magnetic field represented by this equation.

MT6-CRT1: MAGNETIC FIELD EQUATION—CURRENT AND THE MAGNETIC FIELD DIAGRAM

The equation below represents the magnetic field at a particular point due to a long, straight current-carrying wire.

$$1\times10^{-5}T = \frac{\left(4\pi\times10^{-7}m\cdot T/A\right)\left(2.50A\right)}{2\pi\left(0.05m\right)}$$

Draw an appropriate diagram of the current-carrying wire and magnetic field represented by this equation. Indicate the point for which the calculation applies.

MT11-CRT1: MOVING RECTANGULAR LOOPS IN UNIFORM MAGNETIC FIELDS—MAGNETIC FLUX

A rectangular wire loop is being pushed to the right at a constant speed into a region in which there is a uniform magnetic field pointing into the page. The figures show the loop at two different times during the process.

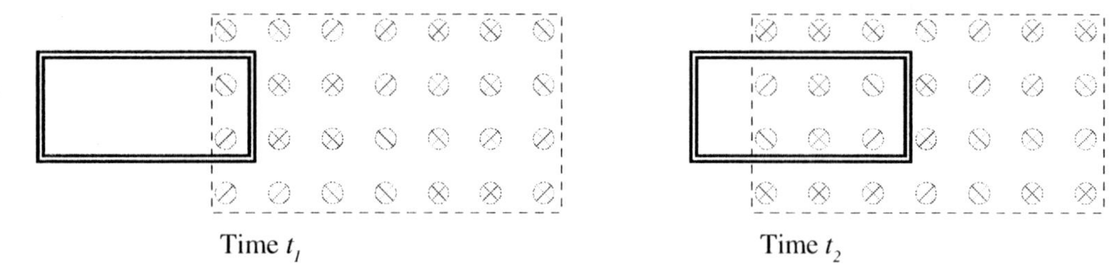

Graph the magnetic flux through the rectangular wire loop as a function of time for the interval between the two times t_1 and t_2.

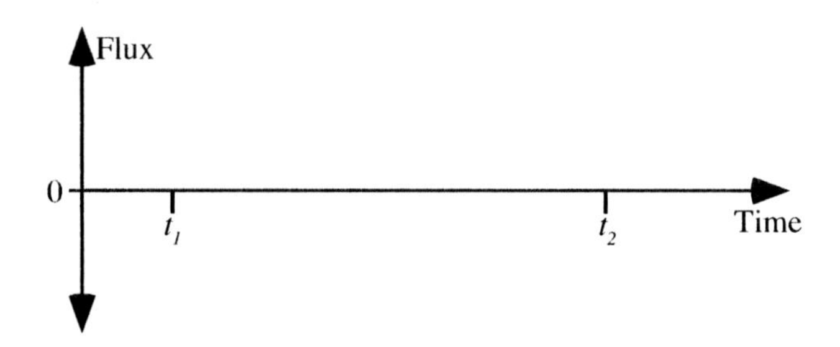

MT11-CRT2: Moving Parallelogram Loop in Uniform Magnetic Fields—Magnetic Flux

A wire loop shaped as a parallelogram is being pulled at a constant speed into a region in which there is a uniform magnetic field pointing into the page.

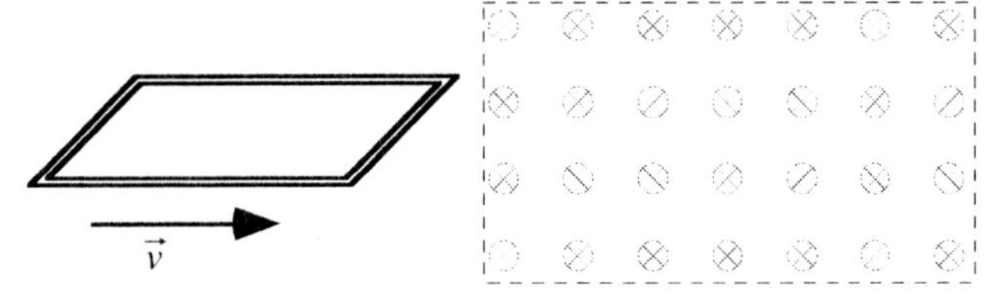

Graph the magnetic flux in the loop versus time as the loop moves into, through, and out of the region.

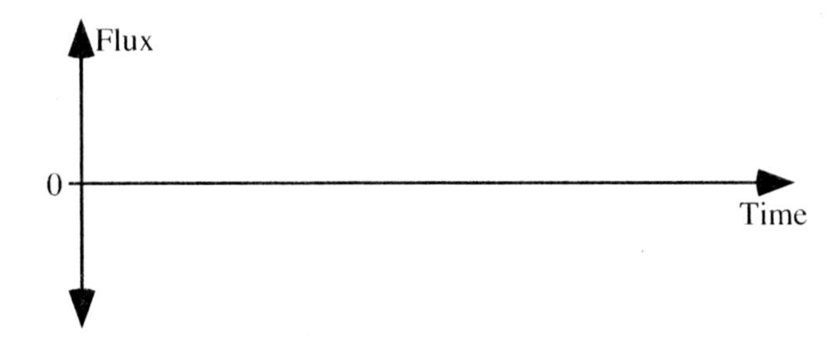

MT12-CRT1: Magnetic Flux vs Time Graph—Emf vs Time Graph

The magnetic flux of a conducting loop that is moving into, through, and out of a region in which there is a uniform magnetic field is shown in the upper graph.

For the time period shown, draw the corresponding emf versus time in the lower graph.

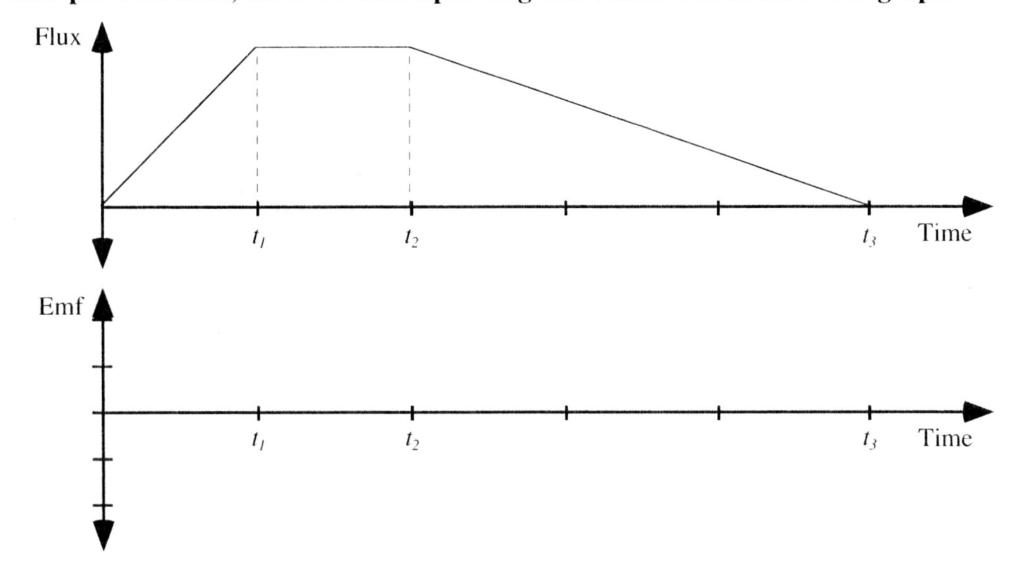

MT13-CRT1: Moving Rectangular Loops in Uniform Magnetic Fields—Current

A rectangular wire loop is being pushed at a constant speed into a region in which there is a uniform magnetic field pointing into the page. The figures show the loop at two different times during the process.

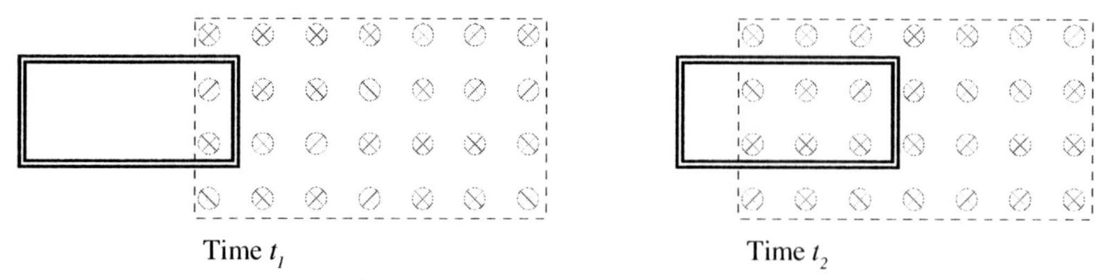

Draw a graph of the current in the rectangular wire loop as a function of time for the interval between the two times shown in the figures.

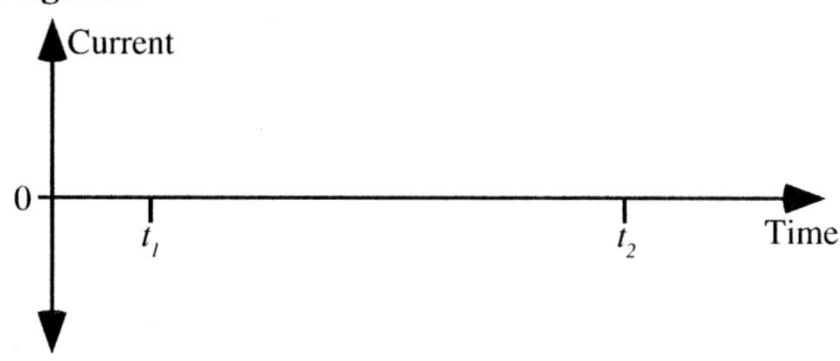

Explain.

BAR CHART TASKS (BCT)

MT5-BCT1: TWO LONG STRAIGHT WIRES—FORCE

Two current-carrying wires are placed side-by-side. The current in the top wire is I and the current in the bottom wire is $3I$.

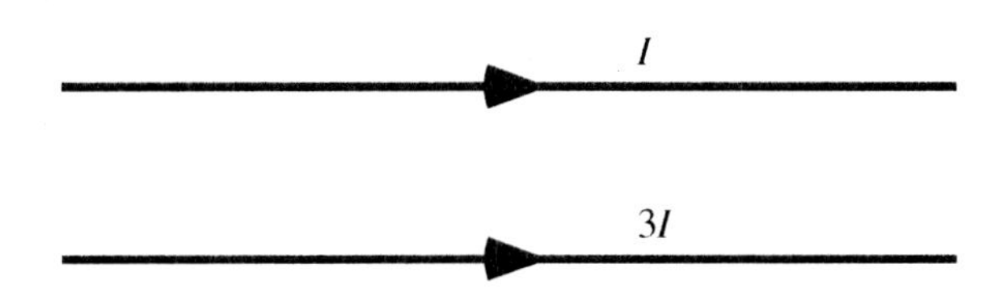

Below is a bar chart showing the magnitude of the magnetic force due to the current in the top wire that is acting on the bottom wire.

Complete the bar chart to show the magnitude of the magnetic force on the top wire due to the current in the bottom wire.

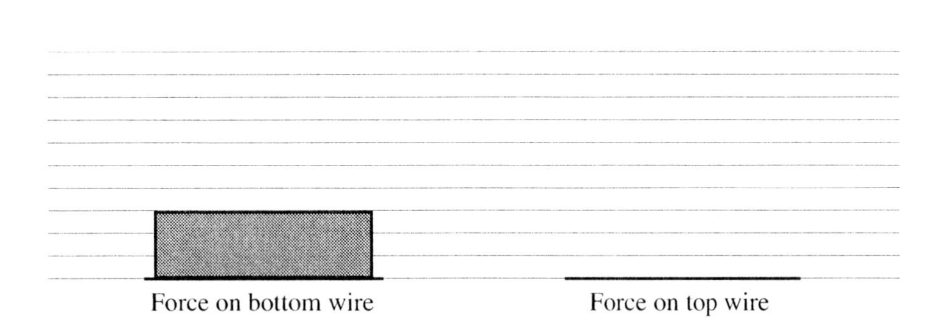

Carefully explain your reasoning.

MT5-BCT2: LONG STRAIGHT WIRE AND RECTANGULAR COIL—FORCE

A rectangular loop of current-carrying wire is placed beside a long straight current-carrying wire. The current in the coil is counterclockwise when viewed as shown. The current in the straight wire is I and the current in the loop is $2I$.

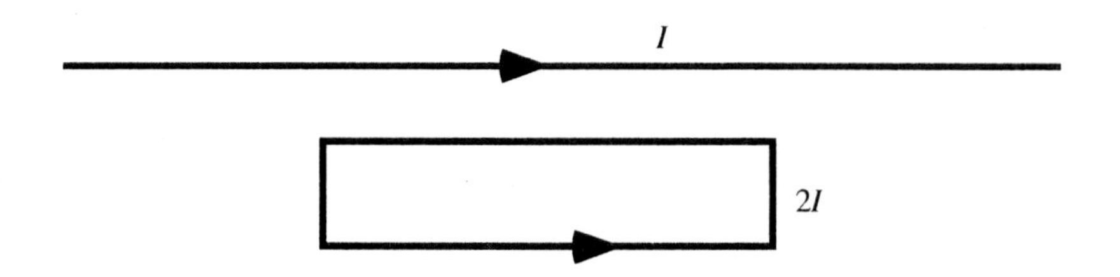

Below is a bar chart showing the magnitude of the magnetic force on the rectangular loop due to the current in the wire.

Complete the bar chart to show the magnitude of the magnetic force on the wire due to the current in the loop.

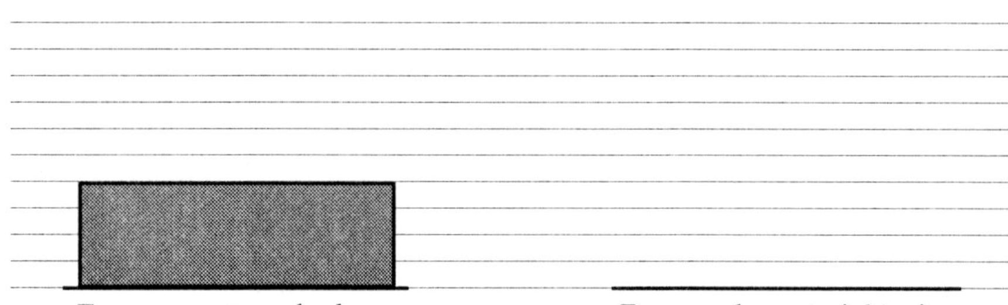

Carefully explain your reasoning.

MT6-BCT1: STRAIGHT CURRENT–CARRYING WIRE—MAGNETIC FIELD

A long, straight conducting wire has a current in the $+x$-direction.

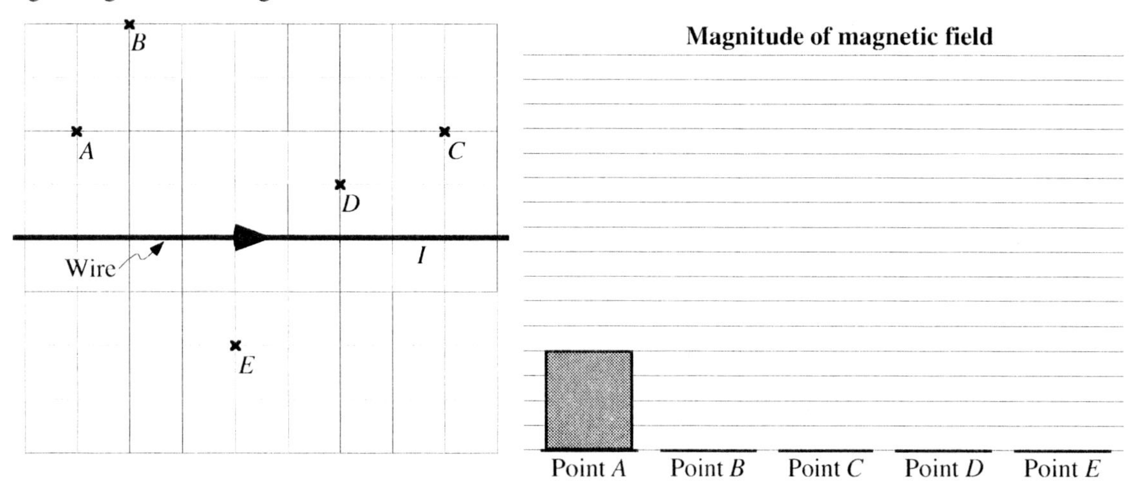

Show the magnitude of the magnetic field at the various points shown in the diagram on the bar chart to the right above. The magnitude at point A is given. **Carefully explain your reasoning.**

MT8-BCT1: THREE PARALLEL CURRENT–CARRYING WIRES I—MAGNETIC FIELD

The bar chart below shows the magnitude of the total magnetic field at point R midway between wires X and Y. The currents in the three wires have the same magnitude.

Complete the bar chart below to show the magnitude of the total magnetic field at point S midway between wires Y and Z.

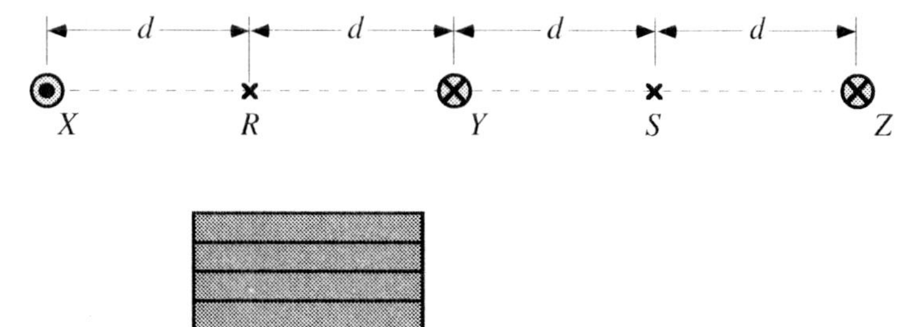

$\vec{\mathbf{B}}$ at point R $\vec{\mathbf{B}}$ at point S

Carefully explain your reasoning.

MT10-BCT1: MOVING CHARGE IN A UNIFORM MAGNETIC FIELD—WORK AND KINETIC ENERGY

A proton moving at a constant velocity is entering a region where there is a uniform magnetic field.

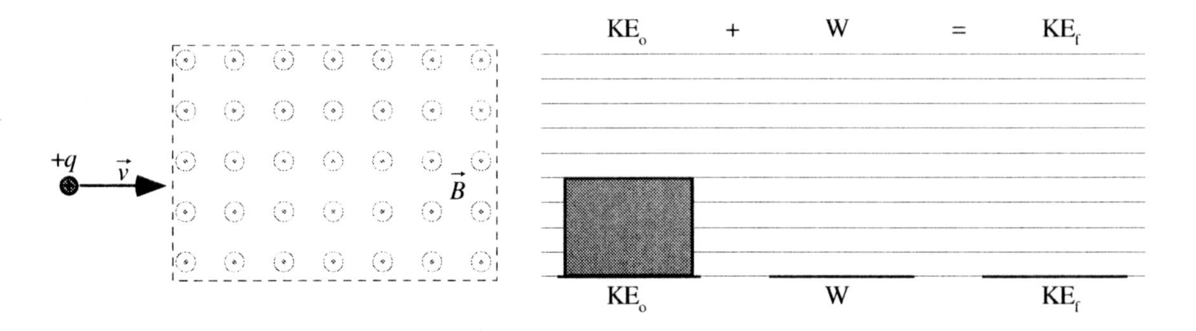

Complete the work-energy bar chart at the right above for the proton from the time just before it enters the magnetic field to the time it leaves the magnetic field. Carefully explain your reasoning.

MT10-BCT2: MOVING CHARGE IN A UNIFORM MAGNETIC FIELD—WORK AND KINETIC ENERGY

An electron moving at a constant velocity is entering a region where there is a uniform magnetic field.

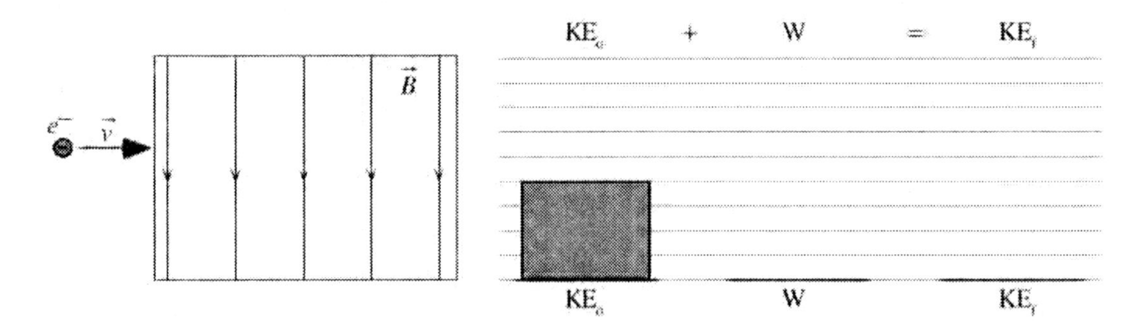

Complete the work-energy bar chart at the right above for the electron from the time just before it enters the magnetic field to the time it leaves the magnetic field. Carefully explain your reasoning.

What, if anything, is Wrong Tasks (WWT)

MT1-WWT1: Electric Charge near a Bar Magnet—Force Direction

A student makes the following statement:

"A positively charged particle placed at rest at the position shown in the figure below will start moving to the right because it will be repelled by the north pole of the magnet."

What, if anything, is wrong with this statement? If something is wrong, explain the error and how to correct it. If the statement is valid, explain why.

MT2-WWT1: Moving Charge within a Uniform Magnetic Field—Force Direction

A student makes the following statement:

"If a proton moving to the east enters a magnetic field pointed away from the center of the Earth, then the direction of the magnetic force on the proton is north."

What, if anything, is wrong with this statement? If something is wrong, explain the error and how to correct it. If the statement is valid, explain why.

MT2-WWT2: Charged Particles and a Uniform Magnetic Field—Direction of Motion

Two particles that have the same mass and electric charge enter the same uniform magnetic field traveling at the same speed. The distance between the two particles is so great that they do not affect each other.

A student makes the following statement:

"These particles will travel in circular paths of equal radius."

What, if anything, is wrong with this statement? If something is wrong, explain the error and how to correct it. If the statement is valid, explain why.

MT3-WWT1: MOVING CHARGE NEAR A STRAIGHT CURRENT–CARRYING WIRE—FORCE

At the instant shown, a positively charged particle has a velocity that is parallel to a current-carrying wire.

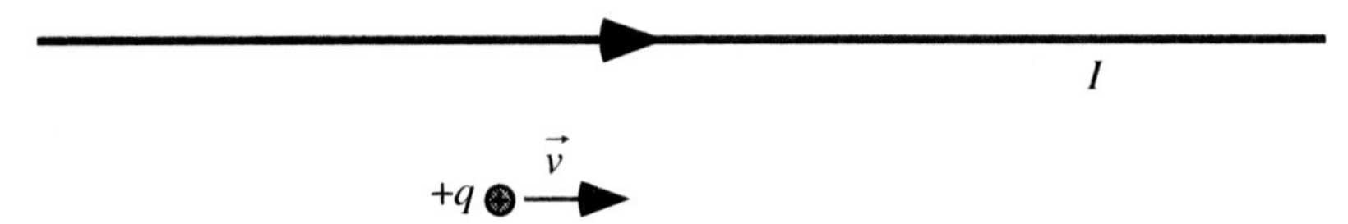

A student makes the following statement:

"The force on the charged particle by the magnetic field is zero because the velocity is parallel to the current in the wire."

What, if anything, is wrong with this statement? If something is wrong, explain the error and how to correct it. If the statement is valid, explain why.

MT4-WWT1: CURRENT–CARRYING WIRE IN A UNIFORM MAGNETIC FIELD—FORCE DIRECTION

A long straight wire is conducting a current whose direction is pointed out of the paper towards you. A student makes the following statement:

"If this current-carrying wire encounters a magnetic field pointing toward the left, then a magnetic force acts on the wire towards the top of the page."

What, if anything, is wrong with this statement? If something is wrong, explain the error and how to correct it. If the statement is valid, explain why.

MT6-WWT1: CURRENT–CARRYING WIRE I—MAGNETIC FIELD DIRECTION

A long straight wire is conducting a current whose direction is pointed out of the paper towards you. A student makes the following statement:

"The magnetic field generated by this wire points straight out from the wire."

What, if anything, is wrong with this statement? If something is wrong, explain the error and how to correct it. If the statement is valid, explain why.

MT9-WWT1: THREE PARALLEL CURRENT–CARRYING WIRES I—MAGNETIC FORCE

A student makes the following statement:

"For three, long, straight parallel wires carrying equal magnitude electric currents located on the same plane, it is not possible to have the total magnetic force at the location of the center wire be zero under any circumstances."

What, if anything, is wrong with this statement? If something is wrong, explain the error and how to correct it. If the statement is valid, explain why.

MT13-WWT1: CHANGING CURRENT—BULB BRIGHTNESS

Two circular loops of wire with small bulbs in them are placed beside two long straight current-carrying wires. In both cases, these loops are the same distance away from the current-carrying wires. The wire loops, bulbs, and long straight wires are identical for the two cases, but the bulb in Case B is dimmer than the bulb in Case A.

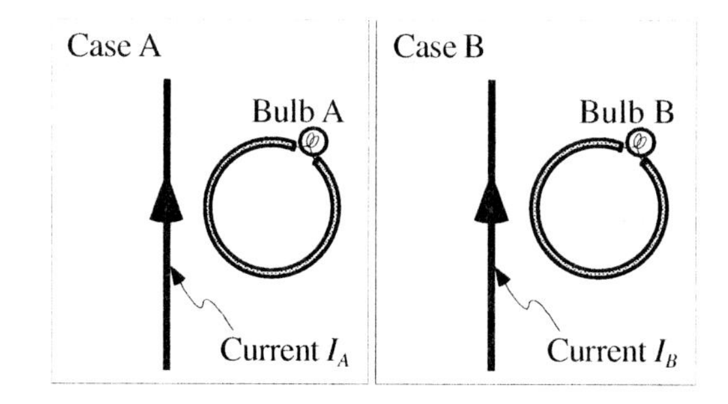

A student makes the following statement:

"At this instant, the current in the wire next to the brighter bulb has the larger value."

What, if anything, is wrong with this statement? If something is wrong, explain the error and how to correct it. If the statement is valid, explain why.

MT14-WWT1: Moving Loop in Uniform Magnetic Field—Induced Magnetic Field

A rectangular wire loop is moving to the right at a constant velocity into a region in which there is a uniform external magnetic field pointing into the page.

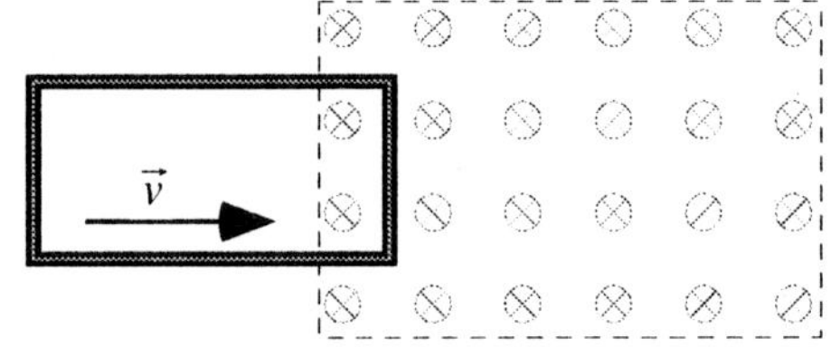

Uniform external magnetic field

A student makes the following statement:

"If a rectangular wire loop is pulled at a constant speed into a region where there is a uniform external magnetic field, the induced magnetic field will increase as more of the loop gets into the field."

What, if anything, is wrong with this statement? If something is wrong, explain the error and how to correct it. If the statement is valid, explain why.

MT14-WWT2: Loop Moving into a Uniform Magnetic Field—Induced Magnetic Field

A rectangular wire loop is moving to the right at a constant velocity into a region in which there is a uniform external magnetic field pointing into the page. Labels A and B in the figure indicate points that are the same distance from the loop.

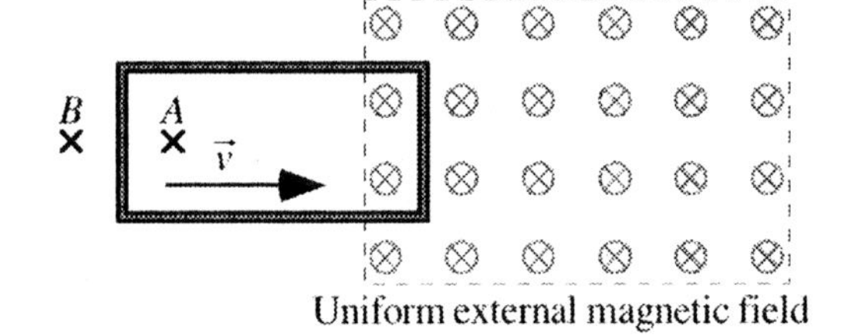

Uniform external magnetic field

A student makes the following statement about the instant shown:

"I think the magnetic field at points A and B will be the same but in opposite directions since they are the same distance from the wire loop."

What, if anything, is wrong with this statement? If something is wrong, explain the error and how to correct it. If the statement is valid, explain why.

TROUBLESHOOTING TASKS (TT)

MT2-TT1: PATH OF A MOVING ELECTRON IN A UNIFORM MAGNETIC FIELD

An electron is moving to the right at a velocity v when it enters a region containing a magnetic field. The external magnetic field is uniform and into the paper. The path of the electron in the magnetic field is indicated by the dotted curve.

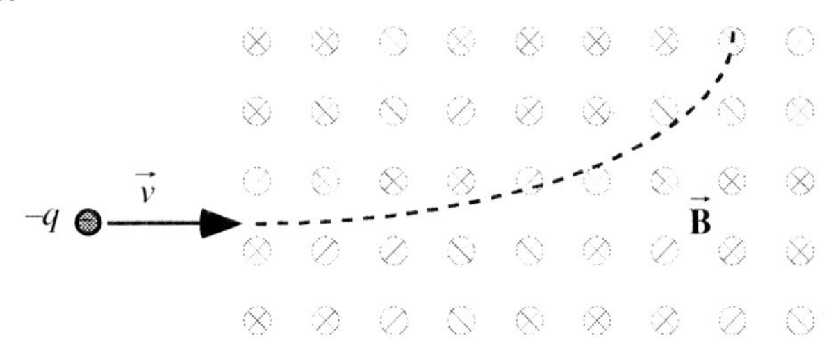

There is at least one error in the diagram. **Identify all errors and explain how to correct them.**

MT3-TT1: MOVING POSITIVE CHARGE NEAR A CURRENT–CARRYING WIRE—FORCE

At the instant shown, a positively charged particle has a velocity that is perpendicular to a current-carrying wire.

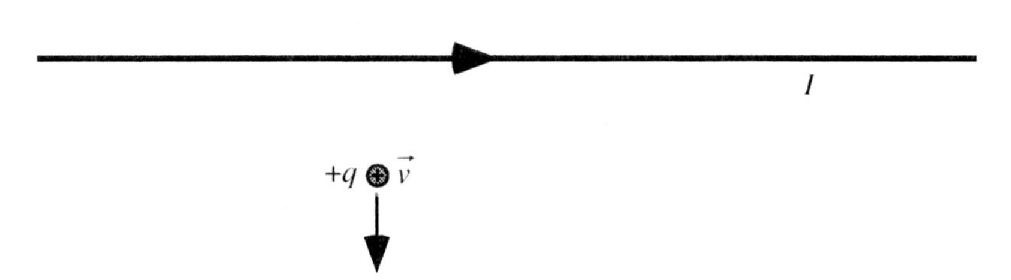

A student makes the following statement:

"The force exerted on the charged particle by the magnetic field is zero because the velocity is parallel to the magnetic field produced by the wire."

There is at least one problem with this student's contention. **Identify any problems and explain how to correct them.**

MT4-TT1: CURRENT–CARRYING WIRE IN A UNIFORM MAGNETIC FIELD—FORCE

A straight wire has a current to the right through a magnetic field. The external magnetic field is uniform and into the paper. A student draws the force on the wire due to the magnetic field as shown.

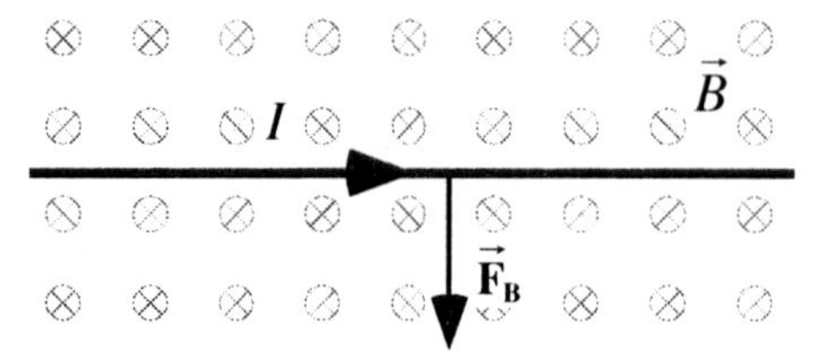

There is at least one problem with the diagram. **Identify any problems and explain how to correct them.**

MT6-TT1: CURRENT–CARRYING WIRE—MAGNETIC FIELD

A student draws the following diagram representing the magnetic field generated by a current-carrying wire.

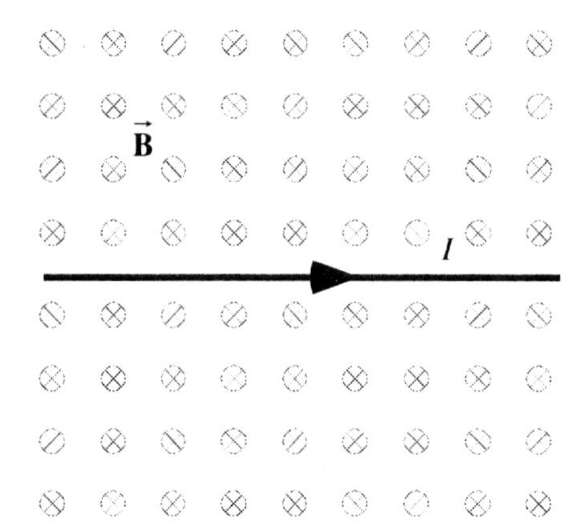

There is at least one problem with the diagram. **Identify any problems and explain how to correct them.**

MT7-TT1: CURRENT–CARRYING CIRCULAR LOOP—MAGNETIC FIELD

A student draws the following diagram representing the magnetic field generated by a current-carrying circular loop of wire.

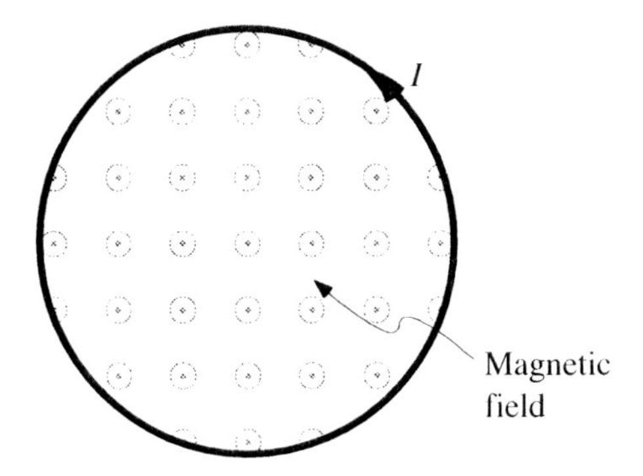

There is at least one problem with the diagram. **Identify any problems and explain how to correct them.**

MT11-TT1: MOVING RECTANGULAR LOOPS IN UNIFORM MAGNETIC FIELDS—MAGNETIC FLUX CHANGE

A student makes the following contention about a situation:

"If a rectangular wire loop is pulled into a region where there is a uniform magnetic field, the rate of change of the magnetic flux will increase as more of the loop gets into the field since there will be more magnetic flux inside the loop."

There is at least one problem with the student's contention. **Identify any problems and explain how to correct them.**

MT12-TT1: MOVING RECTANGULAR LOOPS IN UNIFORM MAGNETIC FIELDS—VOLTAGE

A rectangular loop with a very small resistor (not shown) is moved into, through, and out of a uniform magnetic field at a constant rate. The plane of the loop is perpendicular to the magnetic field.

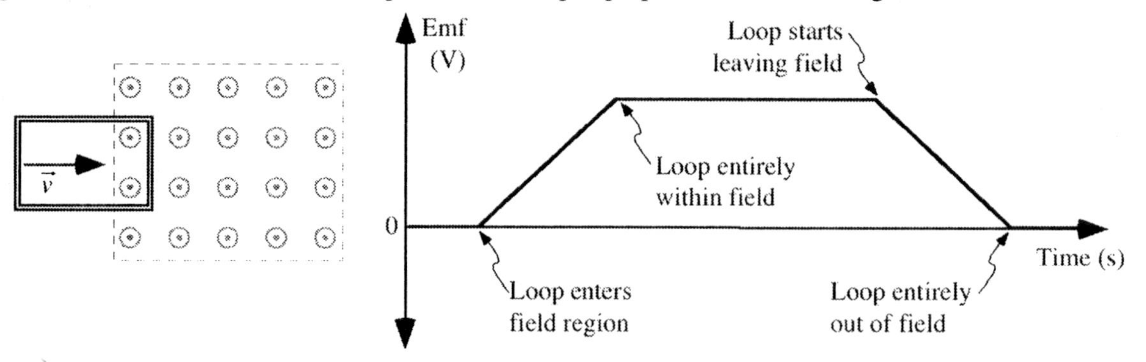

A student draws the graph at right above of emf across the resistor versus time.

There is at least one problem with the student's graph. **Identify any problems and explain how to correct them.**

MT13-TT1: CHANGING CURRENT—BULB BRIGHTNESS

Two circular loops of wire with small bulbs in them are sitting beside two long straight current-carrying wires. In both cases, these loops are the same distance away from their respective current-carrying wires. The bulbs are both lit but differ in brightness. The wire loops, bulbs, and long straight wires are identical for the two situations.

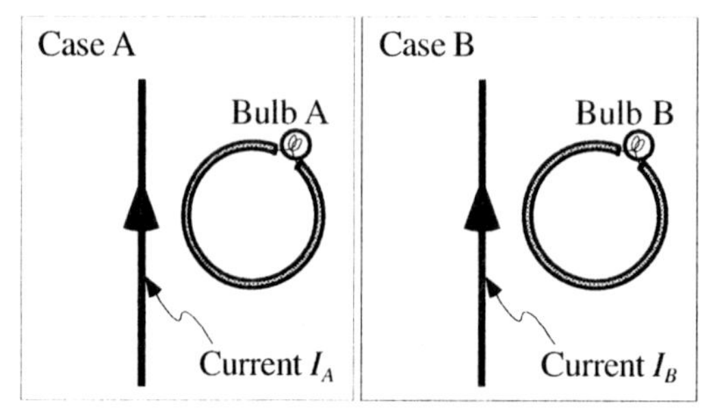

A student makes the following contention:

"The long wire next to the brighter bulb has the larger current in it."

There is at least one problem with the student's contention. **Identify any problems and explain how to correct them.**

MT14-TT1: MOVING LOOPS IN UNIFORM MAGNETIC FIELDS—MAGNETIC FIELD

A vertical conducting bar forms a complete rectangular loop with the conductor it is sliding on. The vertical bar is moving to the right at a constant speed in a region with a uniform external magnetic field. Points A and B are the same distance from the sliding bar at the instant shown.

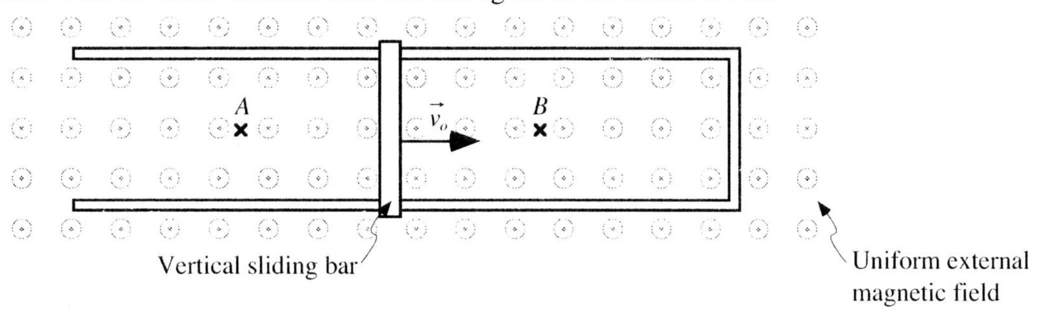

A student makes the following contention about this situation:

"The total magnetic field at point A is the same as the external field and at point B it will be larger."

There is at least one problem with the student's contention. **Identify any problems and explain how to correct them.**

PREDICT AND EXPLAIN TASKS (PET)

MT2-PET1: ELECTRON MOVING INTO A UNIFORM MAGNETIC FIELD—ELECTRON

An electron moving at a constant speed toward the east enters a region in which the magnetic field is uniform and is directed towards the north.

What will happen to the electron when it enters the magnetic field? Explain.

MT2-PET2: PROTON AT REST IN A UNIFORM MAGNETIC FIELD—PROTON

A proton is placed at rest in a region in which the magnetic field is uniform and is directed toward the north.

What will happen to the proton when it is released? Explain.

MT2-PET3: PROTON MOVING INTO A UNIFORM MAGNETIC FIELD—PROTON

A proton moving at a constant speed toward the east enters a region in which the magnetic field is uniform and is directed toward the east.

What will happen to the proton when it enters the magnetic field? Explain.

MT8-PET1: THREE PARALLEL CURRENT–CARRYING WIRES I—CHANGE S[illegible]E CURRENT

At a point that is midway between wires X and Y, the total magnetic field is direc[illegible]toward the top of the page. All wires have the same magnitude of current. The distance between wires X [illegible] Y is the same distance as between Y and Z.

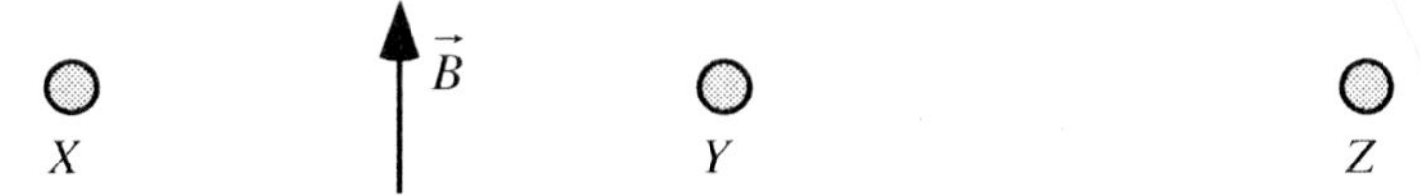

How could a single current be changed to produce a total field at this point in the op[illegible]ite direction? Explain.

MT10-PET1: MOVING CHARGE IN A UNIFORM MAGNETIC FIELD—KINETIC ENERGY

A negatively charged particle is projected into a region in which there is a uniform magnetic field that is constant over time. The particle's initial velocity is perpendicular to the field.

What will happen to the kinetic energy of the particle and why?

MT13-PET1: CIRCULAR LOOPS WITHIN A SOLENOID—AMMETER

A solenoid with a large diameter has a current I. A circular wire loop of smaller diameter is placed inside the solenoid with the axis of the loop parallel to the axis of the solenoid. The wire loop is connected to an ammeter. The current in the solenoid increases to three times its initial value during a 50-millisecond time interval.

What will we observe on the ammeter? Explain.

WORKING BACKWARDS TASKS (WBT)

MT2-WBT1: EQUATION FOR A CHARGE AND A MAGNETIC FIELD I—PHYSICAL SITUATION

Draw and describe a physical arrangement for which the equation below could apply.

$$-6.49\times10^{-12}\,\hat{\mathbf{i}}\,\mathrm{N} = (-1.602\times10^{-19}\,\mathrm{C})(2.50\times10^{7}\,\hat{\mathbf{j}}\ \mathrm{m/s})\ \times\ (1.62\hat{\mathbf{k}}\ \mathrm{T})$$

MT2-WBT2: EQUATION FOR A CHARGE AND A MAGNETIC FIELD II—PHYSICAL SITUATION

Draw and describe a physical arrangement to which the equation below could apply.

$$r = \frac{(1.67\times10^{-27}\,\mathrm{kg})\ (5.00\times10^{6}\,\mathrm{m/s})}{(1.602\times10^{-19}\,\mathrm{C})\ (0.200\ \mathrm{T})}.$$

MT2-WBT3: PROTON MOVING STRAIGHT THROUGH MAGNETIC FIELD—CAUSE

A small region of space is shown that has a uniform magnetic field coming out of the page. A uniform electric field exists in the same region, but is not shown. A proton enters this region with a speed v_0. It travels through this region in a straight line.

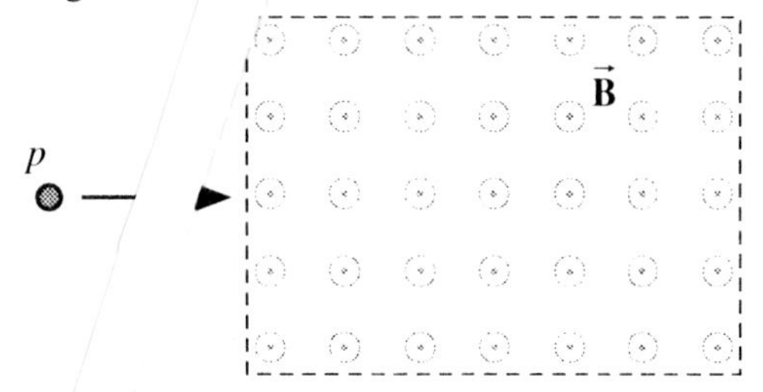

Describe the electric field in this region.

What would happen if the proton was replaced by an electron traveling at the same speed v_0?

What would happen if the proton was initially traveling faster than v_0?

MT4-WBT1: EQUATION FOR A CURRENT AND A MAGNETIC FIELD II—PHYSICAL SITUATION

Draw and describe a physical arrangement to which the equation below could apply.

$$0.720 \text{ N} = (3.20 \text{ A})\,(1.50 \text{ m})\,(0.300 \text{ T}) \sin 30^\circ$$

MT7-WBT1: EQUATION FOR A CURRENT AND A MAGNETIC FIELD—PHYSICAL SITUATION

Draw and describe a physical arrangement for which the equation below could apply.

$$6.37\text{A} = \frac{\left(1.60\times 10^{-5}\,\text{T}\right)\left(0.025\text{m}\right)}{10\left(4\pi\times 10^{-7}\,\text{T}\cdot\text{m/A}\right)}$$

MT8-WBT1: EQUATION FOR THREE CURRENTS—PHYSICAL SITUATION

Draw and describe a physical situation to which the following could apply.

$$\left|\vec{B}\right| = \frac{\mu_0(2\text{ A})}{2\pi(4\text{ cm})} + \frac{\mu_0(4\text{ A})}{2\pi(8\text{ cm})} - \frac{\mu_0(2\text{ A})}{2\pi(4\text{ cm})}$$

MT8-WBT2: Three Parallel Current–Carrying Wires I—Direction of Currents

The arrow in the figure below represents the direction of the total magnetic field at a point midway between wires *X* and *Y*. Wire *Z* at right has a current into the page. All three wires are carrying the same magnitude of current.

For the situation shown, what can you determine about the directions of the currents in wires *X* and *Y*?

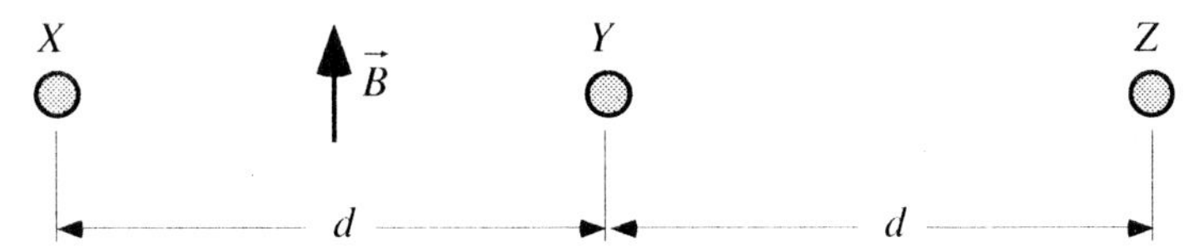

Carefully explain your reasoning.

MT8-WBT3: Three Parallel Current–Carrying Wires II—Direction of Currents

The arrow shows the total magnetic field that is acting on wire *Y*. Wire *Z* at right has a current flowing into the page.

For the situation shown, what can you determine about the directions of the currents in wires *X* and *Y*?

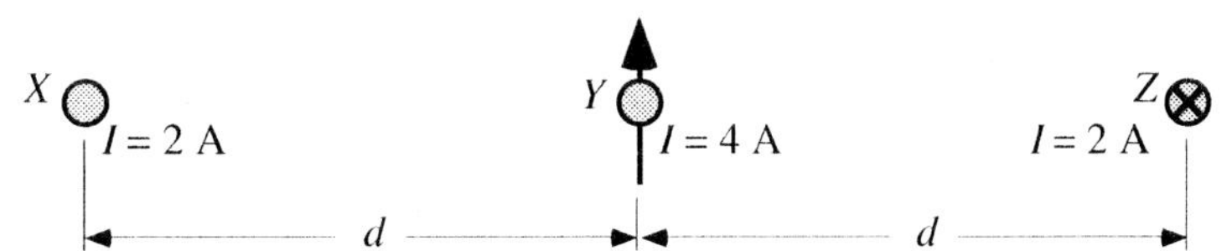

Carefully explain your reasoning.

MT9-WBT1: Three Parallel Current–Carrying Wires I—Direction of Currents

The arrow in the diagram shows the direction of the magnetic force on wire B due to wires A and C. Wire C at right has a current into the page.

For the situation shown below, what can we determine about the directions of the currents in wires A and B?

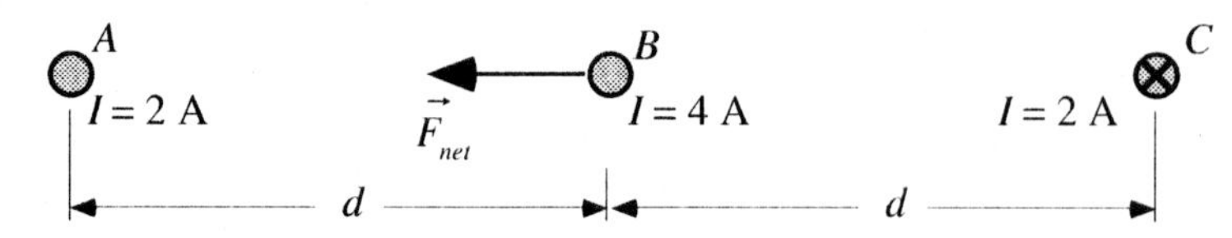

Carefully explain your reasoning.

MT10-WBT1: Charge and a Magnetic Field—Physical Situation

A charged particle moves into a region in which there is a magnetic field. A particle's kinetic energy-versus-time graph for the time that the particle is in the field region is given below.

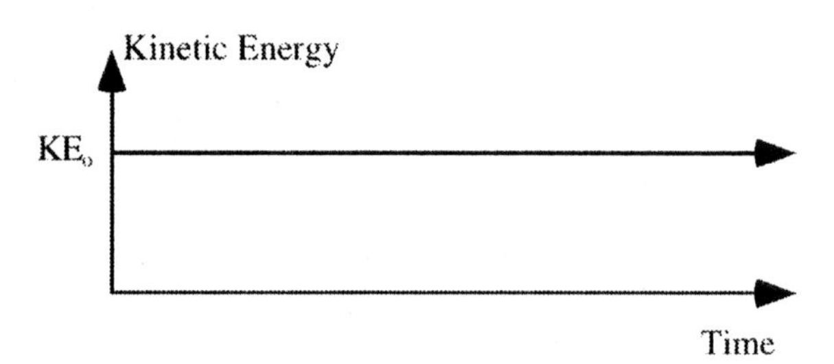

Describe a physical situation that corresponds to this graph.

MT11-WBT1: Moving Rectangular Loops in Uniform Magnetic Fields—Situation

Describe a physical situation, involving a rectangular wire loop and a uniform magnetic field, to which the graph of magnetic flux versus time pictured below could apply.

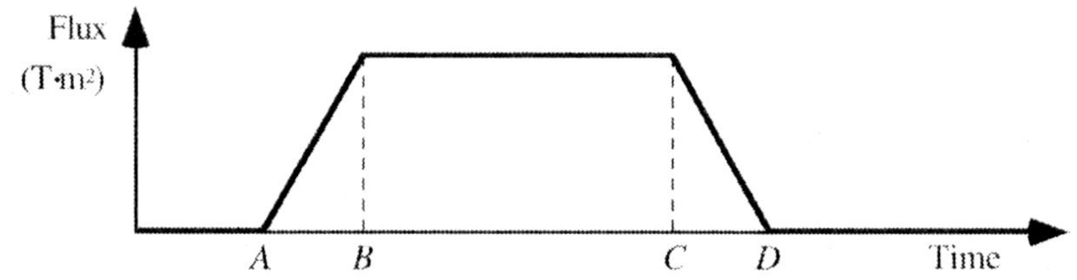

Explain and describe the events at the labeled times.

MT11-WBT2: Magnetic Flux versus Time Graph—Loop Characteristics

A rectangular wire loop passes through a region containing a uniform magnetic field pointing into the page. The region is 160 centimeters long and 100 centimeters high, and the magnetic field is 0.1 Tesla. A graph of magnetic flux versus time for the rectangular wire loop is pictured below.

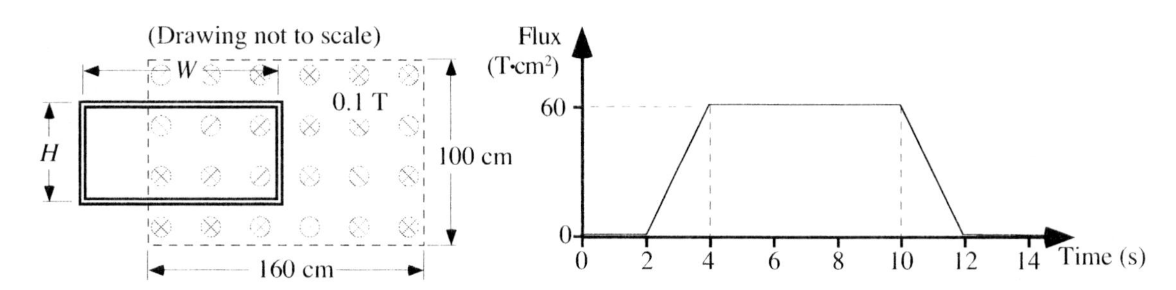

1) What is the speed of the rectangular wire loop? Explain.

2) What is the width W of the rectangular wire loop? Explain.

3) What is the height H of the rectangular wire loop? Explain.

MT11-WBT3: MOVING RECTANGULAR LOOPS IN UNIFORM MAGNETIC FIELDS—SITUATION

Describe a physical situation, involving a rectangular wire loop and a uniform magnetic field, to which the graph of the change in magnetic flux versus time below could apply.

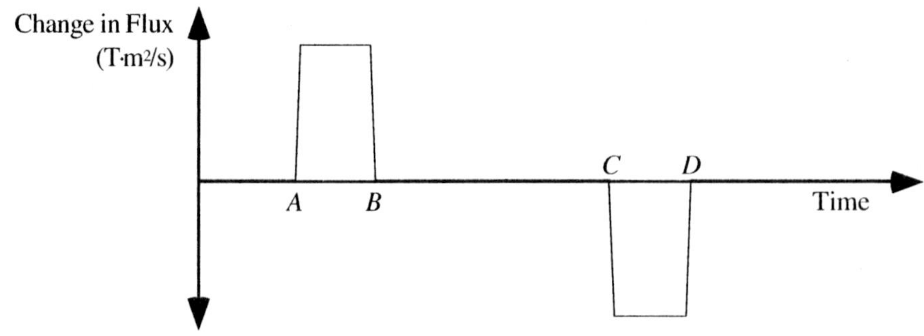

Explain and describe the events at the labeled times.

MT12-WBT1: MOVING RECTANGULAR LOOPS IN UNIFORM MAGNETIC FIELDS—SITUATION

Describe a physical situation, involving a rectangular loop of wire and a uniform magnetic field, to which the graph of emf versus time below could apply.

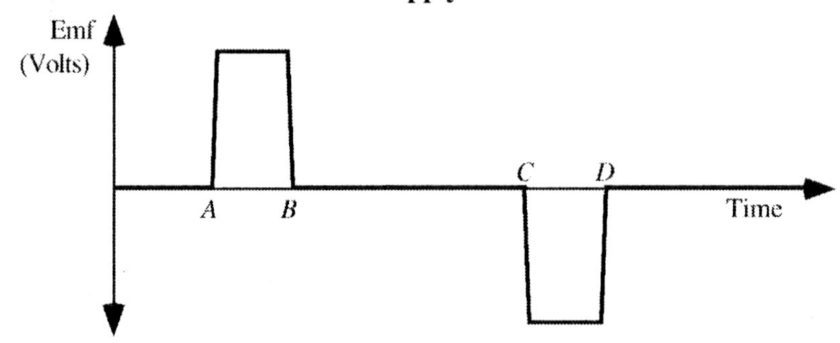

Explain and describe the events at the labeled points.